INTERNATIONAL WILDLIFE ENCYCLOPEDIA

THIRD EDITION

Volume 17

SEA–SOL

Marshall Cavendish Corporation
99 White Plains Road
Tarrytown, New York 10591–9001

Website: www.marshallcavendish.com

© 2002 Marshall Cavendish Corporation

Library of Congress Cataloging-in-Publication Data

Burton, Maurice, 1898-
    International wildlife encyclopedia / [Maurice Burton, Robert Burton] .-- 3rd ed.
        p. cm.
    Includes bibliographical references (p.).
    Contents: v. 1. Aardvark - barnacle goose -- v. 2. Barn owl - brow-antlered deer -- v. 3. Brown bear - cheetah -- v. 4. Chickaree - crabs -- v. 5. Crab spider - ducks and geese -- v. 6. Dugong - flounder -- v. 7. Flowerpecker - golden mole -- v. 8. Golden oriole - hartebeest -- v. 9. Harvesting ant - jackal -- v. 10. Jackdaw - lemur -- v. 11. Leopard - marten -- v. 12. Martial eagle - needlefish -- v. 13. Newt - paradise fish -- v. 14. Paradoxical frog - poorwill -- v. 15. Porbeagle - rice rat -- v. 16. Rifleman - sea slug -- v. 17. Sea snake - sole -- v. 18. Solenodon - swan -- v. 19. Sweetfish - tree snake -- v. 20. Tree squirrel - water spider -- v. 21. Water vole - zorille -- v. 22. Index volume.
    ISBN 0-7614-7266-5 (set) -- ISBN 0-7614-7267-3 (v. 1) -- ISBN 0-7614-7268-1 (v. 2) -- ISBN 0-7614-7269-X (v. 3) -- ISBN 0-7614-7270-3 (v. 4) -- ISBN 0-7614-7271-1 (v. 5) -- ISBN 0-7614-7272-X (v. 6) -- ISBN 0-7614-7273-8 (v. 7) -- ISBN 0-7614-7274-6 (v. 8) -- ISBN 0-7614-7275-4 (v. 9) -- ISBN 0-7614-7276-2 (v. 10) -- ISBN 0-7614-7277-0 (v. 11) -- ISBN 0-7614-7278-9 (v. 12) -- ISBN 0-7614-7279-7 (v. 13) -- ISBN 0-7614-7280-0 (v. 14) -- ISBN 0-7614-7281-9 (v. 15) -- ISBN 0-7614-7282-7 (v. 16) -- ISBN 0-7614-7283-5 (v. 17) -- ISBN 0-7614-7284-3 (v. 18) -- ISBN 0-7614-7285-1 (v. 19) -- ISBN 0-7614-7286-X (v. 20) -- ISBN 0-7614-7287-8 (v. 21) -- ISBN 0-7614-7288-6 (v. 22)
    1. Zoology -- Dictionaries.   I. Burton, Robert, 1941-   . II. Title.

    QL9 .B796 2002
    590'.3--dc21

                                                2001017458

Printed in Malaysia
Bound in the United States of America

07 06 05 04 03 02 01   8 7 6 5 4 3 2 1

Brown Partworks

Project editor: Ben Hoare
Associate editors: Lesley Campbell-Wright, Rob Dimery, Robert Houston, Jane Lanigan, Sally McFall, Chris Marshall, Paul Thompson, Matthew D. S. Turner
Managing editor: Tim Cooke
Designer: Paul Griffin
Picture researchers: Brenda Clynch, Becky Cox
Illustrators: Ian Lycett, Catherine Ward
Indexer: Kay Ollerenshaw

Marshall Cavendish Corporation
Editorial director: Paul Bernabeo

Authors and Consultants

Dr. Roger Avery, BSc, PhD (University of Bristol)

Rob Cave, BA (University of Plymouth)

Fergus Collins, BA (University of Liverpool)

Dr. Julia J. Day, BSc (University of Bristol), PhD (University of London)

Tom Day, BA, MA (University of Cambridge), MSc (University of Southampton)

Bridget Giles, BA (University of London)

Leon Gray, BSc (University of London)

Tim Harris, BSc (University of Reading)

Richard Hoey, BSc, MPhil (University of Manchester), MSc (University of London)

Dr. Terry J. Holt, BSc, PhD (University of Liverpool)

Dr. Robert D. Houston, BA, MA (University of Oxford), PhD (University of Bristol)

Steve Hurley, BSc (University of London), MRes (University of York)

Tom Jackson, BSc (University of Bristol)

E. Vicky Jenkins, BSc (University of Edinburgh), MSc (University of Aberdeen)

Dr. Jamie McDonald, BSc (University of York), PhD (University of Birmingham)

Dr. Robbie A. McDonald, BSc (University of St. Andrews), PhD (University of Bristol)

Dr. James W. R. Martin, BSc (University of Leeds), PhD (University of Bristol)

Dr. Tabetha Newman, BSc, PhD (University of Bristol)

Dr. J. Pimenta, BSc (University of London), PhD (University of Bristol)

Dr. Kieren Pitts, BSc, MSc (University of Exeter), PhD (University of Bristol)

Dr. Stephen J. Rossiter, BSc (University of Sussex), PhD (University of Bristol)

Dr. Sugoto Roy, PhD (University of Bristol)

Dr. Adrian Seymour, BSc, PhD (University of Bristol)

Dr. Salma H. A. Shalla, BSc, MSc, PhD (Suez Canal University, Egypt)

Dr. S. Stefanni, PhD (University of Bristol)

Steve Swaby, BA (University of Exeter)

Matthew D. S. Turner, BA (University of Loughborough), FZSL (Fellow of the Zoological Society of London)

Alastair Ward, BSc (University of Glasgow), MRes (University of York)

Dr. Michael J. Weedon, BSc, MSc, PhD (University of Bristol)

Alwyne Wheeler, former Head of the Fish Section, Natural History Museum, London

Picture Credits

Heather Angel: 2310, 2371, 2383, 2384, 2399; **Ardea London Ltd:** Francois Gohier 2357, P. Morris 2412; **Neil Bowman:** 2335, 2363, 2428; **Bruce Coleman:** Franco Banfi 2312, 2365, 2366, 2440, Erwin & Peggy Bauer 2352, E. Bjurstrom 2390, Dr. Hermann Brehm 2387, Fred Bruemmer 2427, Jane Burton 2314, 2353, 2354, 2376, John Cancalosi 2420, M.P.L. Fogden 2408, 2389, 2402, Christer Fredriksson 2316, 2432, Bob Glover 2341, 2396, 2397, Charles & Sandra Hood 2315, 2364, 2423, Janos Jurka 2345, 2391, 2407, Steven C. Kaufman 2323, P. Kaya 2419, Gunter Kohler 2317, Gordon Langsbury 2338, 2362, Werner Layer 2351, 2400, 2409, Robert Maier 2406, Antonio Manzanares 2343, Joe McDonald 2418, 2421, 2422, Rine Van Meurs 2337, Pacific Stock 2329, Dr. Eckart Pott 2318, 2344, Andrew Purcell 2348, Hans Reinhard 2433, Kim Taylor 2355, Colin Varndell 2392, Jorg & Petra Wegner 2321, 2322, 2375, Staffan Widstrand 2401, Rod Williams 2342, 2347, 2434, Gunter Ziesler 2361; **Corbis:** Hal Beral 2367, Gary Braasch 2411, Ralph Clevenger 2424, Clive Druett/Papilio 2426, Bates Littlehales 2333, George McCarthy 2346, Joe McDonald 2404, Mary Ann McDonald 2443, David A. Northcott 2405, 2416, Lynda Richardson 2381, 2382, 2425, Lawson Wood 2379, Robert Yin 2368, 2444; **Chris Gomersall:** 2334, 2386, 2388, 2429; **NHPA:** Agence Nature 2311, 2327, A.N.T. 2330, 2380, 2435, Joe Blossom 2442, G.J. Cambridge 2398, Stephen Dalton 2377, 2378, Ken Griffiths 2417, Dan Griggs 2360, E. Hanumantha ROA 2403, Martin Harvey 2325, 2326, John Hayward 2359, Daniel Heuclin 2308, 2370, 2373, Image Quest 3-D 2355, Hellio & Van Ingen 2438, 2439, T. Kitchin & V. Hurst 2393, 2394, 2395, Yves Lanceau 2445, Andy Rouse 2324, 2358, Norbert Wu 2385, 2415; **Oxford Scientific Films:** L.M. Crowhurst 2339, Terry Heathcote 2374, Paul Kay 2349, Breck P. Kent 2332, Rudie Kuiter 2430, John McCammon 2350, TC Nature 2340, Tony Tilford 2319; **Still Pictures:** Fred Bavendam 2441, Nigel J. Dennis 2372, Serge Dumont 2331, Michel Gunther 2369, Yves Lefevre 2309, Thomas D. Mangelsen 2437, Roland Seitre 2436. **Artwork:** Ian Lycett 2320, 2336, Catherine Ward 2356, 2401, 2414, 2431.

Contents

SEA SNAKE

SEA SNAKES ARE REPTILES that have returned to the sea from the land and have mostly become fully adapted for life in the water. They differ from land snakes in having the body flattened from side to side, with a flattened, paddle-shaped tail that is used for propulsion.

There are two main groups of sea snakes. "True" sea snakes are believed to have evolved from land-living, venomous snakes from Australia, and are still placed in the same family as cobras by some zoologists. There are about 50 species. They are highly adapted to the marine environment, having nostrils located on the top of the snout so that they can breathe when most of the head is under water. Females produce well developed young without laying external eggs, so they do not need to return to land to breed. The nostrils can be closed by valves that prevent water from entering when the snake dives. As in most snakes there is only one lung, and this extends for most of the length of the body to provide buoyancy. Glands near the base of the tongue excrete salt that the snake takes in from the surrounding sea, and the snakes can breathe to a certain extent through their skins, although they cannot entirely rely on this method. Though they swim gracefully when they are in the water, true sea snakes are almost helpless on land because the scales on their undersides are very small and do not provide grip. They never leave the water voluntarily.

The other group is called the sea kraits, and its members are less fully adapted for life in the sea. They are similar to the land kraits, genus *Bungarus*, also related to cobras. They too have compressed tails like true sea snakes, but the scales on the under surface are large, so they can move on land. There are six species, five of which live around coral reefs where they feed on eels, but they also spend a lot of time out of the water on small coral islands. The sixth species, the Rennell Island sea krait, *L. crockeri*, lives in a small, landlocked lake in the Solomon Islands.

True sea snakes are highly venomous, with the venom glands located just below the eye, and are a particular danger to fishers. Sea kraits are also venomous, but they very rarely bite, so they are not regarded as dangerous.

Sea wanderers

Sea snakes live in a variety of habitats. Most of them live in relatively shallow coastal waters; they are especially common around coral reefs. Some are associated with the muddy water around mangrove trees. Other lesser-known species live in deeper water.

The brightly colored yellow-bellied sea snake is one of the few sea snakes with a common name. Most of the body is yellow, with a dappled tail and a wide, dark stripe that runs along the body and covers the top of the head. This is a pelagic (oceanic) snake: it wanders long distances out to sea. It may be found anywhere in tropical seas from the east coast of Africa, through the Indian and Pacific Oceans to the west coasts of the Americas. This is the largest geographical distribution of any living reptile.

Of the other species of sea snakes, some have a spine on the snout, while others have thornlike scales above each eye. The color also varies. Some species, especially those living among waving fronds of seaweed, have dark, transverse bars on a pale background.

The cobralike fangs of sea snakes are hollow and not very large. They are permanently erect and situated at the front of the mouth. Sea snakes usually hunt at night, striking swiftly at their prey. Their powerful venom kills their victim within moments, preventing the victim's escape into crevices in the coral. Some species feed exclusively on eels, others on small fish. Some small fish have the habit of congregating

Laticauda colubrina, called the giant banded sea krait or the wide-faced sea krait, is widely distributed off the coasts of India, Indonesia, New Guinea and islands in the western Pacific. It is not devoted to the marine life, and must return to land to lay its eggs, hence another common name, the amphibious sea snake.

SEA SNAKES

CLASS	**Reptilia**
ORDER	**Squamata**
SUBORDER	**Serpentes**
FAMILY	**Elapidae**

SUBFAMILY **Hydrophiinae (true sea snakes); Laticaudinae (sea kraits)**

GENUS AND SPECIES **Hydrophiinae: *Pelamis platurus*, yellow-bellied sea snake; *Aipysurus laevis*, olive sea snake; *Hydrophis inornatus*. Laticaudinae: *Laticauda colubrina*, others**

LENGTH
Usually 2–3 feet (60–90 cm)

DISTINCTIVE FEATURES
Paddle-shaped tail (both families); small scales on underside (true sea snakes); large scales on underside (sea kraits)

DIET
Fish, particularly eels

BREEDING
Age at first breeding (*A. laevis*): female 3 years, male 5 years; number of young: typically 2 to 6 (true sea snakes)

LIFE SPAN
True sea snakes: at least 10 years

HABITAT
Most species: shallow coastal waters in the Tropics; sea kraits: especially coral reefs and atolls

DISTRIBUTION
Most species found from Horn of Africa, through Southeast Asia to north Taiwan and coasts of Australia; yellow-bellied sea snake occurs anywhere in tropical and subtropical regions of Indian and Pacific Oceans

STATUS
Some species abundant; others unknown

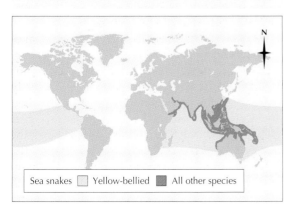

Sea snakes | Yellow-bellied | All other species

under any floating object, such as a mass of seaweed or a piece of wood, and consequently a sea snake, floating quietly at the surface, automatically attracts its natural food.

Massing at the surface

Incredible numbers of sea snakes are seen from time to time massed together on the surface of the sea. There is one record from the Malacca Strait, for instance, of a 10-foot (3-m) wide belt of sea snakes writhing at the surface, extending for at least 60 miles (100 km). These enormous concentrations may be connected with some phase of the breeding cycle. At this time the snakes also seem particularly aggressive.

Many sea snakes, especially in the Philippines, have a limited economic value. Some are hunted for their skins, from which a good leather is made, whereas others are bottled and sold as an aphrodisiac and as a kind of medical cure-all. The sea krait *Laticauda semifasciata* was either used for its skin or was kept alive in sacks and shipped to Japan and the Ryukyu Islands, where it was smoked and eaten as a delicacy. This fishing still goes on but to a limited degree.

Sea snakes are typically associated with coral reefs, where they tend to specialize in hunting slender fish such as eels and gobies.

SEA SPIDER

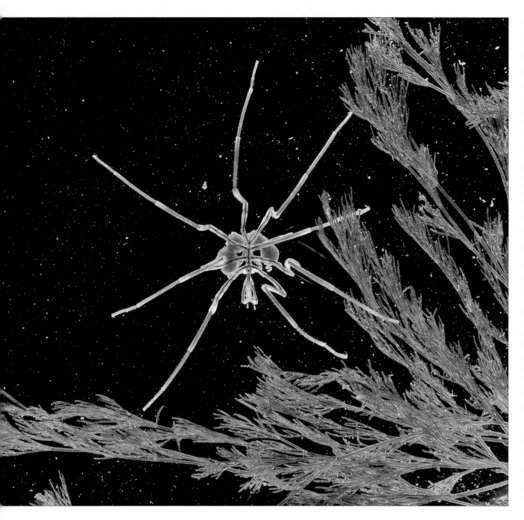

the other end is a tiny abdomen, reduced to a little nodule. At the front of the head are feeding appendages called chelicerae and pedipalps, thought to be related to the feeding appendages of spiders and scorpions. However, sea spiders also have an unusual feeding organ, a proboscis, with a large, triangular sucking mouth at the end. The mouth can actually be larger than the diameter of the body. Farther back on the head of the male is a pair of unique appendages that hang downward and backward. They are called ovigers or ovigerous legs, and are used for holding and brooding the eggs laid by the female. In both sexes there are two or four simple eyes on a small turret, located where the head joins the first segment of the body. The smallest sea spiders are only a fraction of an inch (less than 1 cm) across. The largest, living in the deep seas, are 2 feet (60 cm) across. They are usually colored cream, yellow or brown.

Sea spiders are found in all seas, from shallow waters, sometimes between tidemarks on rocky coasts, down to at least 23,000 feet (7,000 m). They are most numerous in polar waters, in both the Arctic and Antarctic.

A male shallow-water sea spider, Nymphon gracile, carries a burden of egg clusters with his specially adapted legs. Also visible are his feeding appendages. The main trunk of the body is short, and no thicker than one of his legs.

SEA SPIDERS, OR PYCNOGONIDS as they are known to zoologists, are also known as Pantopoda, which means "all legs." Their enormously long legs so dwarf the rest of the body that there is no room in the trunk for most of the internal organs. Remarkably, both the digestive system and the reproductive system extend well into the animal's legs. Sea spiders are not closely related to land spiders, but both belong to the phylum Arthropoda, which also includes the insects, crustaceans and other animals with jointed legs. The enigmatic sea spiders bear such a number of unique and unusual features that some zoologists have placed them in their own subphylum, emphasising how different they are from other arthropods.

Unusually in arthropods, there is no consistency among the sea spiders in number of walking legs. Some species have five pairs, others six pairs, but most species have four pairs. The walking legs are attached to the segmented trunk which is extremely reduced and slender. At the front of the trunk is a simple head, and at

The shore and the deep sea

Sea spiders spend their lives clinging to sea firs (hydroids), anemones and other sedentary (usually stationary) animals, swaying gently with the movements of the surrounding water. They seem to be able to walk over such animals without being harmed by their stinging cells. Those living on the shore are usually found sheltering under stones or pebbles when the tide is out. Shore-living sea spiders are the best studied species, but they are in the minority. Most of the over 600 species live far from the tidal zone, at great depths, and are very difficult to observe in their natural habitat.

Rasping and sucking

Sea spiders feed on coral polyps, sea firs, sea anemones, jellyfish and sea cucumbers. They pierce the skin of these animals by rasping at it using three teeth on the proboscis, then they insert the proboscis and suck up the body fluids of their prey. Some sea spiders are also thought to feed on microscopic algae. Sea spiders have a

SEA SPIDERS

PHYLUM	**Arthropoda**
CLASS	**Pycnogonida**
ORDER	**Pantopoda**

GENUS AND SPECIES **Many, including**
Nymphon gracilis, Collessendeis colossea
and *Pycnogonum* spp.

LENGTH
Across body and legs: ⅛–24 in. (0.3–60 cm)

DISTINCTIVE FEATURES
Ovigers (leglike appendages); chelicerae (feeding appendages); proboscis; slender trunk; simple head; 3 thoracic segments; small abdomen; 4 to 6 pairs of walking legs

DIET
Benthic (bottom-dwelling) invertebrates: hydroids and polyzoans, coral polyps, sea firs, sea anemones, jellyfish, sea cucumbers

BREEDING
Sexes separate; male tends eggs. Typical protonymphon (larva) undergoes 8 instars before adulthood: early instars are legless, later instars have 3 pairs of walking legs, final stages have at least 4 pairs.

LIFE SPAN
Not known

HABITAT
Rocky shores, where they cling to hydroids or bryozoans; also seabeds in deep water

DISTRIBUTION
Worldwide

STATUS
Not known

■ *Pycnogonum stearnsi* □ *Acehlia australiensis*

A rare picture of the giant deep-sea sea spider, Collessendeis colossea, *proboscis turned toward the camera. Before pictures like this, scientists thought this sea spider used its spindly legs to hold itself clear of the soft benthic ooze over which it walks.*

Brooding appendages

The male sea spider fertilizes the female's eggs as she lays them, then gathers them into a rounded mass using his specialized ovigers. He glues the eggs in place with cement excreted from organs in the fourth joints of some of the legs. In some species the male carries one ball of eggs on each oviger; in others he may have half a dozen. From the eggs hatch larvae known as protonymphons, which have no trace of walking legs. They can cling by their mouthparts, and for a while they either hold onto their father or are parasitic on or inside coral polyps, jellyfish, sea slugs, clams and sea cucumbers. Later, each protonymphon grows three pairs of walking legs and finally changes to the adult with at least four pairs.

Deep sea "private eye"

Much of our knowledge of sea spiders comes from individuals living on larger animals brought up from the seabed in nets. Large sea spiders, such as *Colossendeis colossea*, are in poor condition by the time the net reaches the surface. Also, pictures from submersible craft that explore the depths of the ocean can now show deep-sea sea spiders in their natural habitat. The bed of the deep ocean is called the benthic zone, or benthos, and features soft mud or ooze consisting of the remains of marine organisms, particularly microscopic diatoms. The pictures show the surface of the ooze marked with tracks from the legs of giant sea spiders, from which we can learn not only about sea spiders but also about the ooze itself. Old ideas about the ooze being very soft, ideas fostered by the stiltlike appearance of many benthic organisms, have been modified by observations of sea spiders moving lightly over the ooze, making only a slight impression with the tips of their legs.

digestive tube from which branch diverticula, or blind tubes, running into each of the four pairs of legs. The reproductive organs are also in the bases of the legs. Thus, the tiny trunk of the animal need not accommodate most of the vital organs other than a simple, tubular heart.

SEA SQUIRT

Not all sea squirts are pale and translucent. These colorful **Polycarpa aureta** are found in the coral reefs of the South Pacific.

SEA SQUIRTS LIVE PERMANENTLY fixed to rocks on the seabed, or to any solid object in the sea that provides enough anchorage. When touched they contract, suddenly squirting out two jets of water. Some sea squirts are solitary, but others live in groups, some of which are closely integrated. The body is enclosed in a firm jellylike tunic, from which sea squirts derive the alternative name, tunicates. The jets of water we see when we touch a sea squirt are ejected through two openings, one at the top of the animal, the inhalant opening, and the other on its side, the exhalant opening. Water is drawn in through the inhalant opening, bringing food and oxygen, and it is driven out through the exhalant opening, carrying away waste products.

Inside the tunic are the vital organs. The inhalant current is drawn into a capacious gullet, which leads into a short intestine that ends near the exhalant opening. Except for one side of the gullet, these organs lie free in the body cavity, or atrium. The walls of the atrium form a muscular layer known as the mantle, which is enclosed by the jellylike tunic. The walls of the gullet are perforated by a thousand or more small vertical slits arranged in rows. Fine blood vessels run through the bars separating these slits, so the walls of the gullet act as gills. The bars are coated with cilia, which, by their concerted beating, set up a current that draws water in through the inhalant opening and drives it through all the

slits. Lying in the atrium is a simple heart from which blood vessels run to the various organs. Those going to the wall of the gullet break up into a fine network to run through the numerous bars separating the gill slits. A small knot of nerve cells, the nerve ganglion, lies in the mantle between the inhalant and exhalant openings and serves as the sea squirt's brain.

Fouling buoys and ships

Sea squirts are found in all seas, from between tidemarks, where they can be numerous on pier piles, to the depths of the ocean. They can even colonize the brackish water of estuaries. The solitary forms hang from the undersides of rocky overhangs, and are found on the hulls of boats left anchored for a long time and on the submerged parts of buoys, pontoon piers and wharves. Those that grow in clusters, as well as some of the solitary forms, grow on rocks, pebbles, bivalve shells or the backs of large crabs. Many species are partially buried in sand or have incorporated small pebbles or sand grains into the outer layers of their tunics. The group-living forms encrust the surfaces of rocks or other hard surfaces, and scores or hundreds of individuals are enclosed in a common tunic. The individuals may be grouped in circles, ovals or stars, as in *Botryllus*. Each individual has its own inhalant opening, but all members of a group share one exhalant opening at the center. In very deep water, down to 16,000 feet (4,900 m) or more, sea squirts settle on manganese nodules or other solid objects, or else have long stalks with the lower part of the stalk buried in the mud, as in the Abyssascidia group.

Cleansing the sea

Together with sponges and bivalve mollusks, sea squirts are living pumping stations, constantly passing water through their bodies and extracting small particles of food from the water. Sea squirts living on the shore and in shallow waters feed on microscopic fragments formed from the breakdown of dead plants and animals, as well as very small members of the living phytoplankton (planktonic plants). Sea squirts living farther from the coasts, where the water is cleaner, feed mainly on plankton, but those living in deep waters rely on dead material, because living phytoplankton are not found below the surface waters. The method of trapping the food, however, is the same in all types. The cilia on the bars between the gill slits waft the particles to one side, where there is a groove bearing cilia and glandular cells giving out mucus that traps the particles. The cilia then

SEA SQUIRT

PHYLUM	**Chordata**
SUBPHYLUM	**Urochordata**
CLASS	**Ascidiacea**
ORDER	**Enterogona**
FAMILY	**Cionidae**
GENUS AND SPECIES	*Ciona intestinalis*

LENGTH
Up to 6 in. (15 cm)

DISTINCTIVE FEATURES
Body soft, translucent and greenish, sometimes with orange or pink stripes; two obvious siphons, one inhalant, one exhalant

DIET
Dead organic debris; phytoplankton

BREEDING
Hermaphroditic. Eggs and sperm released in large numbers. Breeding season: summer (cold and temperate waters), year round (Tropics); larval period: minutes or hours.

LIFE SPAN
1–2 years

HABITAT
Attached to rocks, shells, seaweeds, corals, piers and boats, down to at least 1,600 ft. (490 m); common in harbors

DISTRIBUTION
Virtually worldwide

STATUS
Abundant

Sea peach, *Tethyum pyriforme*
Lightbulb sea squirt, *Clavelina lepadiformis*

drive the food-laden mucus up toward the top of the gullet, from where it is passed down another groove into the stomach.

Tadpole larvae

Each sea squirt is hermaphoditic (both male and female). It sheds its eggs and sperm into the water, where the eggs are fertilized by sperm from another sea squirt. In some species the eggs stay in the parent body and are fertilized by sperm carried in on the inhalant current. In either case the larva swims freely in the sea and looks like a tiny, long-tailed tadpole with a sucker on the front of the head and a mouth just behind this. In the tail is a notochord and a nerve cord, as in young vertebrates, and the internal organs are generally better developed than in the adult. After a while the tadpole settles head-down on a solid surface, absorbs its tail, notochord and nerve cord, and in their place grows two openings, becoming a baby sea squirt.

Real-life Peter Pan

Related to sea squirts are animals known as appendicularians. They are like the tadpole larva of a sea squirt but, like Peter Pan, they never grow up. Instead they become sexually mature as juveniles, and the adult phase of the life history is lost. This kind of reproduction is known as paedogenesis and is similar to neoteny in some amphibians. An appendicularian has a special way of feeding. It secretes mucus over its head that traps small plankton. These are eaten, but small inedible particles also fall onto the mucus and after a while the mucus becomes cluttered with these, so the appendicularian throws off its mucus cap and spends the next half hour secreting a new one.

One of the unusual, free-swimming appendicularians, Fritillaria sp., a relative of the sea squirts. Being pelagic, it can actively feed on plankton, rather than feeding only passively on detritus.

SEA URCHIN

A sea urchin (edible sea urchin, above) moves by extending and contracting its thin, sucker-tipped tube feet.

SEA URCHINS, ALSO KNOWN as sea hedgehogs, are the spiniest of the phylum Echinodermata, or spiny skins. This group includes starfish, brittle stars, sea lilies and sea cucumbers. The sea urchin's internal organs are enclosed in a test, which typically takes the form of a more or less rounded and rigid box made up of chalky plates fitting neatly together. However, some sea urchins have leathery, flexible tests. In sea urchins with a rigid box the shape may be nearly spherical, rounded and somewhat flattened, heart-shaped or like a flattened disk, as in the sand dollar (discussed elsewhere in this encyclopedia). In most sea urchins the test is covered with spines, which may be short and sharp, long and slender or thick and few in number. When the spines are removed, knobs are revealed on some of the plates. These form part of the ball-and-socket joints on which the spines move. Removal of the spines also makes visible the double rows of pinholes arranged in a series of five, forming a star in the heart urchin, order Spatangoidea (so named for its heart-shaped body), or running from the bottom to the top of the test when this is spherical. In the living sea urchin the tube feet project through these pinholes. Among the spines are small jointed rods with two or three jaws at the top that act like tiny pincers. These are known as pedicellariae, and each moves like the spines on a ball-and-socket joint. Living sea urchins may be green, yellow, red, orange or purple in color. The smallest sea urchins are barely ½ inch (1.25 cm) across; the largest are up to 1½ feet (46 cm) across, including the spines.

The 800 species of sea urchins have a worldwide distribution, mainly in shallow waters, usually at less than 600 feet (180 m), although some live down to 1,500 feet (460 m).

Walking on the seabed

More than 2,000 years ago Aristotle wrote of sea urchins' teeth as resembling a horn lantern with the panes of horn left out. There are five vertical teeth supported on a framework of rods and bars, and the whole structure, teeth and supporting skeleton, has subsequently become known as Aristotle's lantern. Some sea urchins move freely over the seabed, or over seaweeds, using their tube feet to pull themselves along. When walking the tube feet are pushed out, their suckers take hold and then the tube feet shorten, pulling the body along. Its course is usually erratic, as first the tube feet on one side are pushed out, then the neighboring tube feet, each side pulling the sea urchin in a slightly different direction. Some sea urchins use their tube feet, assisted by the spines, while others walk on the spines in a deliberate way, pursuing a steady course. Other sea urchins, including the heart urchins, plow through sand and burrow into it, using their spines.

The common heart urchin, *Echinocardium cordatum*, sinks vertically into the sand, to a depth of 8 inches (20 cm), twice its own length or more. It lines its vertical shaft with mucus and pushes several extensible tube feet up the shaft to the surface of the sand to breathe. It uses other tube feet to keep a horizontal shaft open behind it to receive its excrement, while tube feet in front pick up particles of food and pass them to the mouth. As the sea urchin moves forward it abandons the vertical shaft by withdrawing its tube feet, and then pushes them up through the sand to make a new vertical shaft to the surface.

There are sea urchins that burrow into soft rock, using their spines to scrape away the surface, some using their teeth as well. The sea urchin *Echinostrephus molaris,* of the Indian Ocean

EDIBLE SEA URCHIN

PHYLUM	**Echinodermata**
CLASS	**Echinoidea**
ORDER	**Diadematoidea**
FAMILY	**Echinidae**
GENUS AND SPECIES	***Echinus esculentus***

LENGTH
Diameter: up to 7 in. (18 cm)

DISTINCTIVE FEATURES
Fairly large urchin; slightly flattened shape; generally whitish, violet or pale pink; numerous spines growing to about ¾ in. (2 cm)

DIET
Grazes on large seaweeds, bryozoans (marine invertebrates), barnacles, young animals and young seaweeds

BREEDING
Sexes separate, but indistinguishable. Breeding season: March–April, in center of range. Eggs and sperms released into sea, where fertilization occurs. Occasional mass spawning. Pluteus larvae spend several days or weeks in plankton before settling on bottom and developing into adults.

LIFE SPAN
Usually 10–15 years

HABITAT
Rocks, stones and gravelly seabeds; usually at 33–130 ft. (10–40 m), but may occur at 0–3,940 ft. (0–1,200 m).

DISTRIBUTION
Northeastern Atlantic, including North Sea

STATUS
Generally common

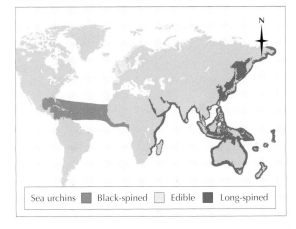

Sea urchins ■ Black-spined □ Edible ■ Long-spined

Eucidaris thouarsii (above) is commonly known as the pencil-spined sea urchin.

and South Pacific, makes cylindrical burrows several inches deep. When feeding, it comes up to the mouth of the burrow. Should anything disturb it, it merely drops into the burrow and wedges itself in with its spines.

On the coast of California, pier piles put in place in the late 1920s and made of steel about 9.5 millimeters thick became perforated by sea urchins of the genus *Strongylocentrotus* after 20 years. Elsewhere the anticorrosive surface layer of the steel piles was abraded and polished by the sea urchins' spines.

Surface-cleaning pincers

Sea urchins are mainly vegetarian, chewing seaweeds with the teeth of their Aristotle's lantern. Burrowing forms eat bits of dead plants in the sand and may also eat animal food. All get some food from cleaning their tests. The pincers of the pedicellariae constantly move around, picking up grains of sand that fall on the skin covering the test as well as any tiny animals that settle, such as barnacle larvae. These are passed from one pedicellaria to another, and on to the mouth.

Free-swimming larvae

Male and female sea urchins shed their sperms and eggs respectively into the sea, where fertilization takes place. The larva, known as a pluteus, is like that of other members of the Echinodermata, with slender arms covered with bands of cilia. Before it settles on the seabed it already has a mouth surrounded by a few tube feet and spines, and the arms are shorter. When the arms are too short for swimming, the tiny sea urchin, barely half-formed, settles on the bottom.

SECRETARY BIRD

Although its hooked bill is similar to that of an eagle, and its long, thin legs recall those of a stork or crane, the secretary bird's relationship to other bird species is uncertain.

from this ragged crest, as the feathers hang down behind the head in the way 18th-century clerks carried their quill pens stuck into their wigs. The species name of secretary birds, *Sagittarius serpentarius*, is derived from their habit of killing snakes. The generic name was inspired by their upright carriage and erect stride, which reminded observers of an archer about to loose an arrow (*Sagittarius* is the Latin for "archer").

Secretary birds live in Africa south of the Sahara, from Senegal and Eritrea to the Cape of Good Hope.

The marching eagle

Secretary birds spend much of their time on the ground, striding about in search of food, and they can run quickly when disturbed. They live in open grassland on plains and savannas or in bush country, provided that the vegetation is not too dense for them to run easily. Secretary birds do not migrate and each usually stays in one place all year round unless there is a food shortage.

Unlike other birds of prey, secretary birds do not have grasping toes. Their toes are stout and blunt and are armed with short, curved talons, so they do not seize their prey with their feet. When they roost in trees at night they squat with their legs doubled under them instead of perching as most other birds do.

Snake stomper

The diet of the secretary bird includes a wide variety of animals, such as large insects, small mammals up to the size of a hare, young birds, snakes, tortoises and lizards. The crop of one secretary bird was found to contain 3 snakes, each over 2 feet (60 cm) long, and 11 small tortoises, each about 2 inches (5 cm) across, together with the remains of lizards, locusts and beetles. These indigestible parts would eventually have been cast out in the form of pellets. The nestlings and eggs of ground-nesting birds, such as plovers, are also taken.

Secretary birds' prowess at killing snakes is well known, and the Afrikaans name for them is *Slangen-vreeter* or snake-eater. They have been seen to kill snakes up to 4 feet (1.2 m) long, and

THE SECRETARY BIRD IS A bird of prey but is unusual in its appearance and habits. It looks rather like a crane, with long legs that are feathered down to the knees. A full-grown male stands about 4 feet (1.2 m) high and has a wingspan of 7 feet (2.1 m). The tail has two very long central feathers, with black spots near the ends. The plumage is gray with black on the wings and legs. On the sides of the face there is a patch of bare red skin, and on the back of the head there is a crest of long, black-tipped feathers. The bird's common name is derived

SECRETARY BIRD

CLASS	**Aves**
ORDER	**Falconiformes**
FAMILY	**Sagittariidae**
GENUS AND SPECIES	***Sagittarius serpentarius***

WEIGHT
5–9½ lb. (2.3–4.3 kg)

LENGTH
**Head to tail: 49–59 in. (125–150 cm);
wingspan: about 6½ ft. (2 m)**

DISTINCTIVE FEATURES
**Very long, partially feathered yellow legs;
hooked bill; loose crest of feathers on nape
of neck; long central tail feathers; hooked
bill; long crest; mainly gray plumage with
black flight feathers**

DIET
**Mostly invertebrates, particularly locusts
and beetles; small mammals; nestlings and
eggs of ground-nesting birds; frogs, lizards,
tortoises, snakes and rodents**

BREEDING
**Age at first breeding: 1 year; breeding
season: varies with wet season; number of
eggs: 2; incubation period: about 42 days;
fledging period: 65–106 days; breeding
interval: 1 year**

LIFE SPAN
Not known

HABITAT
**Savanna with short grass and scattered
acacia trees**

DISTRIBUTION
**Senegal and Gambia east to Ethiopia, south
to South Africa**

STATUS
Widespread but scarce

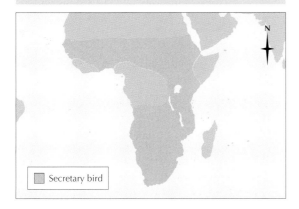

Secretary bird

they are sometimes encouraged to live near farms to help keep down numbers of snakes and other vermin in the locality. If they are confronted with a snake that they cannot subdue on the ground, secretary birds often take to the air with the snake and drop it to the ground from a height.

Secretary birds feed predominantly on small mammals, such as mice and hares, and large insects, particularly beetles and locusts, although the degree to which they are taken varies with locality and individual preference. Some secretary birds prefer to eat tortoises and snakes.

Secretary birds hunt by walking through the grass, occasionally stamping vigorously, apparently to flush prey. Small animals are caught in the bill after a quick lunge with the neck but larger animals are killed by kicking. When dealing with snakes, the secretary bird holds its wings out, using them as shields, for the birds are not immune to snakebites.

Long breeding season
Usually silent, secretary birds emit distinctive groaning and croaking calls during courtship, when members of a pair chase one another in the air or on the ground in a dramatic display. The undulating courtship flight is similar to that of raptors, each bird soaring upward before tipping forward into a steep dive and then ascending again. On the ground, the two birds run around with their wings raised. They may be joined by other secretary birds, who behave in a similar way, although generally a pair of secretary birds will defend its territory against others.

The secretary bird is named for the loose feathers at the back of its head, which bear a passing resemblance to pen quills.

The nest, which is used year after year, is built of sticks in the top of a thorn tree such as an acacia and lined with dry material, mainly tufts of grass. It is flat and measures about 8 feet (2.4 m) across. From the time that the greenish white eggs are laid to the fledging of the chicks, some 4 months elapse, a considerable period, and the chicks stay with the parents for some time after they have learned to feed themselves. The female usually lays two or three eggs (sometimes one or three), which she incubates on her own for a period of more than 9 weeks. During this period she is fed by the male, which brings food to her that he carries in his crop and regurgitates on the nest rim. The eggs are 3 inches (7.5 cm) long, white or pale green splashed with red. At first the nestlings are covered with white down, which later turns gray. The characteristic head feathers sprout after about 3 weeks. Although adults use their feet as weapons, the chicks have short legs with brittle bones.

Within a month of the chicks' birth, the female leaves the nest to forage for food with the male. Both parents bring food to the chicks, which are fed at first by regurgitation of liquid food. Later, the parents simply deposit food in the nest for the chicks to feed themselves. The chicks leave the nest 65–80 days after hatching, at which time they are barely capable of flight. They then spend about a month in the local area, learning how to hunt, before beginning to forage for food with their parents.

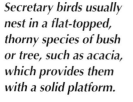

Secretary birds usually nest in a flat-topped, thorny species of bush or tree, such as acacia, which provides them with a solid platform.

Obscure relationships

The secretary bird's relationship to other species of birds remains a subject of scientific debate. The bill is very eaglelike, although not very powerful, and some of the bird's displays are similar to those of eagles. Once airborne, a secretary bird is a good flier and soars easily, but takeoff is accomplished only after a long run with wings outstretched, and the bird generally prefers to walk. It is possible, therefore, that the secretary bird is a relative of eagles and other birds of prey, but is losing the characteristic powerful flight and developing a terrestrial way of life. This explanation has resulted in the creation of a separate family for the secretary bird alone, the Sagittariidae.

However, secretary birds also display similarities to bustards, family Otididae. It is also possible that secretary birds belong to the same order as cranes, the Gruiformes, and may be related to the seriema (discussed elsewhere in this encyclopedia). Relationships between animal groups can be demonstrated by similarity of their blood proteins in much the same way that human relationships are shown by blood groups. Secretary birds show the greatest morphological, behavioral and molecular affinities with the Falconiformes, especially members of the family Accipitridae, which includes the eagles, sparrowhawks and Old World vultures. They also exhibit some behavioral affinities with the Ciconiidae, the storks' family.

SEED SNIPE

THE FOUR SPECIES OF seed snipe are small birds, between 6¼–12 inches (16–30 cm) in length. They are not closely related to snipes at all but were so named because they were thought to eat mostly seeds and because they have the rapid, erratic flight and rasping alarm call of true snipes. Their bodies are plump and their bills are short and conical like those of sparrows. Their legs are short and weak, and when seed snipes alight they drop their bodies to the ground like nightjars or nighthawks, which also have weak legs.

The plumage is generally speckled brown above and white underneath. The least, pygmy or Patagonian seed snipe, *Thinocorus rumicivorus*, is the smallest species, measuring only 6¼–7½ inches (16–19 cm). The male has a characteristic black band around the neck and down the breast. The rufous-bellied or Gay's seed snipe, *Attagis gayi*, is gray above with black markings. The other two species are the white-bellied seed snipe, *A. malouinus*, and the gray-breasted or D'Orbigny's seed snipe, *T. orbignyianus*.

Seed snipe are confined to South America from Ecuador to the archipelago of Tierra del Fuego and Isla de los Estados. The white-bellied seed snipe lives only in southern Chile and Argentina and on the Falkland Islands, one of the regions where the least seed snipe is also found.

Migrate to avoid snow

Seed snipe live as high as the snow line of the Andes, at 18,000 feet (5,500 m), and occur in a range of habitats such as mountain pastures, pampas and the rocky shores of Patagonia. The seed snipe of the higher slopes move down to the valleys at the first sign of snow, whereas those of southern Patagonia migrate north.

Southern populations of least seed snipe are partially migratory. They winter on sandy grasslands as far north as Uruguay. Gray-breasted seed snipe leave Tierra del Fuego and higher parts of the Patagonian Andes for the lower mountain ridges to the east. There are several records of least seed snipe and two of white-bellied seed snipe on the Falkland Islands.

Seed snipe are very reluctant to fly when disturbed, and it is possible to ride or walk through a flock without their flying up. They usually run for a short distance and then conceal themselves by crouching with the head flat on the ground. Although they spend most of their

Least seed snipe nest on the ground, making basic hollows lined with moss and pieces of plants. Their eggs, buff colored with dark speckles, are well camouflaged.

In fresh breeding plumage the male least seed snipe has a striking black stripe around the neck and down the breast.

time on the ground, singing females often use elevated positions such as rocks. In the sheep-ranching areas of Patagonia least seed snipe use fence posts for the same reason.

The name seed snipe is a little misleading because these birds feed mainly by biting off tiny pieces of plants such as buds, leaf tips and small green leaves. The least seed snipe may take more seeds than the other species. Seed snipe are not known to drink in natural conditions, presumably gaining all the moisture they need from the plants they eat and from dew.

Hidden eggs

Seed snipe nest on the ground, the nests grouped as close as 165 feet (50 m). There are four eggs, sometimes three, which are well camouflaged. The least seed snipe nests are crude scrapes in the ground, often lined with mosses and plant debris, and placed close to a stone or dwarf shrub. Many are placed in dried loose dung. The seed snipe displays to distract predators. It jumps in the air and falls on its back, and as the intruder approaches, gets up, runs away and falls over again. Little else is known about the breeding habits of seed snipe except that the chicks leave the nest shortly after hatching.

Outside the breeding season seed snipe live in small flocks, although up to 80 rufous-bellied seed snipe have been seen in one flock. White-bellied seed snipe sometimes congregate in large numbers when the snow is deep. The least seed snipe usually remains in flocks of 10 to 20 throughout the year.

LEAST SEED SNIPE

CLASS	**Aves**
ORDER	**Charadriiformes**
FAMILY	**Thinocoridae**
GENUS AND SPECIES	***Thinocorus rumicivorus***

ALTERNATIVE NAMES
Patgonian seed snipe; pygmy seed snipe

LENGTH
Head to tail: 6¼–7½ in. (16–19 cm)

DISTINCTIVE FEATURES
Plump body; small, conical bill; short legs; wedge-shaped tail. Male: gray face, neck and breast; black borders to white throat and underparts connected by black line down center of breast; dark upperparts. Female: pale buff head, neck and breast; dark border to white throat.

DIET
Mainly buds and leaf tips of herbs; seeds

BREEDING
Age at first breeding: 6 months or more; breeding season: August–February; number of eggs: usually 4; incubation period: about 26 days: fledging period: about 49 days; breeding interval: several broods per year

LIFE SPAN
Not known

HABITAT
Mountain pasture and semidesert with scattered grasses, herbs and succulent plants, up to altitude of 15,000 ft. (4,600 m)

DISTRIBUTION
Summer: southern Ecuador, south through Peru and Chile to southern Argentina. Winter: northern Argentina and Uruguay.

STATUS
Common

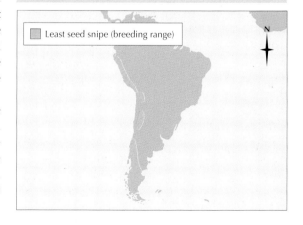

Least seed snipe (breeding range)

SERIEMA

THE SERIEMAS, ALSO CALLED cariamas, are large cranelike birds. The name is derived from the Tupi Indian words for small rhea. Seriemas look rather like long-legged bustards and stand about 2½ feet (75 cm) tall. The body is often held horizontally on a pair of long legs and the tail and neck are also fairly long. The bill is hooked and a bushy crest rises from near the base of the bill.

The red-legged or crested seriema, *Cariama cristata*, has a bushy crest about 4 inches (10 cm) long. Its plumage is generally grayish with fine markings on the body and broad black-and-white bands on the wings and tail. Burmeister's seriema, *Chuna burmeisteri*, is slightly smaller but has a longer and thicker crest and a brownish plumage. Seriemas are confined largely to an inland area of South America, ranging from central Brazil to Uruguay and northern Argentina.

Different habits

Seriemas are very weak fliers. The red-legged seriema lives in open scrub, grassland and lightly wooded areas, spending most of its time on the ground, running when danger threatens but roosting in trees at night. It breeds in captivity and is often kept with domestic chickens to hunt vermin. In common with the secretary bird, seriemas eat snakes, although this reputation is sometimes exaggerated. They also make good watchdogs, because they will scream when an intruder approaches. Burmeister's seriema lives in open woodland and roosts in high trees. It has a more restricted range than the red-legged seriema and is a wary bird, perhaps because it is often hunted. As a result, it is less well known.

Seriemas live in pairs or small flocks and feed on a variety of food, including insects, worms, snails, rodents, reptiles and occasionally fruit. They sometimes kill small snakes, though they are not immune to snake poison.

Courtship displays

The male red-legged seriema displays by stretching his flight feathers and strutting in front of the female with his head down and crest raised. Seriemas are well known for their loud and far-carrying calls. Both sexes help in nest building, which takes about one month. The nest is usually in a tree 3¼–16½ feet (1–5 m) above the ground. It is made of twigs and small branches, lined with mud and leaves.

The two eggs, which differ from each other in the amount of gloss, shape and color, are pink when newly laid but fade to dull white. Burmeister's seriema nests in trees. The red-legged seriema incubates the eggs for 25–30 days and the chicks stay in the nest until they are well grown. They are fed by both parents and leave

The red-legged seriema has a distinctive call that has been likened to maniacal laughter or the sharp yelping of puppies.

In addition to its usual open savanna habitat, the red-legged seriema also lives in human-made grassy areas formerly covered by tropical forest, at elevations of up to 6,500 feet (2,000 m).

the nest aged 12 to 15 days, following their parents in their search for food. They fledge at the age of about 1 month.

Prehistoric ancestors

Extinct ancestors of the seriemas probably include the huge, ostrichlike, flesh-eating birds that survived in South America up to 60 million years ago. These huge birds had stout necks, heavy heads and large hooked bills. Some stood up to 8 feet (2.4 m) tall. It has been suggested that they died out because they could not compete with predatory mammals. An earlier relative, dating from 50 million years ago, was the massive *Diatryma*, fossils of which have been found in the Eocene rocks of North America. It stood 7 feet tall (2.1 m) and had a massive head with a huge puffinlike, hooked bill. Fossils of similar large carnivorous birds exist in Eocene rocks in Britain, Belgium, France and Switzerland.

RED-LEGGED SERIEMA

CLASS	Aves
ORDER	Gruiformes
FAMILY	Cariamidae
GENUS AND SPECIES	*Cariama cristata*

ALTERNATIVE NAMES
Crested seriema; red-legged cariama; crested cariama

WEIGHT
About 3½ lb. (1.5 kg)

LENGTH
Head to tail: up to 3 ft. (90 cm)

DISTINCTIVE FEATURES
Large, long-legged bird; relatively long neck and tail; stout, hooked bill; crest of raised, slightly stiff feathers; mostly grayish with brown markings; white underparts; banded black-and-white flight and tail feathers

DIET
Insects, rodents, lizards, frogs and birds; some snakes, seeds and fruits

BREEDING
Age at first breeding: 8 months in captivity; breeding season: May–September; number of eggs: 2; incubation period: 25–30 days; fledging period: about 30 days; breeding interval: 1 year

LIFE SPAN
Not known

HABITAT
Open savanna and lightly wooded areas; recently colonized humanmade grassy areas formerly covered by tropical forest

DISTRIBUTION
Central and eastern Brazil through eastern Bolivia and Paraguay to Uruguay and central Argentina

STATUS
Common (central Brazil); varies elsewhere

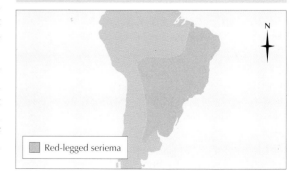

Red-legged seriema

SEROW

THE SEROW IS ONE OF SEVERAL animals known as goat-antelopes, the others including the goral, chamois and Rocky Mountain goat. Although it was previously put in a different genus from the others, the serow is very closely related to the goral, with which it now shares a generic name. The serow differs in being larger, at least 32 inches (81 cm) tall, whereas gorals are never more than 28 inches (71 cm). Serows are also distinguished in having small face glands that give out a duikerlike odor. These glands lie in pits in the skull. The serow's horns are similar to those of the goral, being black, closely ridged and about 8 inches (20 cm) long. They point straight back in line with the face and then curve down very slightly.

Of the three species, the maned, or mainland, serow stands 39–42 inches (99–107 cm) high and weighs 150–200 pounds (68–91 kg). It has a thin rough coat, grizzled black to red, a strong mane on the back of the neck and very long ears. It lives from Sumatra north to the Himalayas, Szechwan and the lowlands of southern China, and varies with its geographic range. The red form is common in the hills of Sumatra and Peninsular Malaysia, but in the lowlands in these areas it is black. Himalayan races have limbs that are chestnut above and dirty white below, and the general coat color is black. Chinese races are black, but the mane is usually white and very long.

The other two are island species. The Japanese serow has a longer, more woolly coat and shorter ears, no mane and a bushy tail; it is black to dark brown but lighter in summer. It is only 32 inches (81 cm) high and weighs 45 pounds (20 kg). The hair may be 4 inches (10 cm) long on the body. The Taiwanese serow is as small as the Japanese species, but its chin and throat are reddish brown, there are white patches on the chest and belly and the hind legs are reddish brown with a black line down the front.

Mountaineering goat-antelopes

Serow, like goral, live in rocky places and are good climbers. They are, however, less agile than the goral, being larger, and they prefer a damper environment. Both are found in the Szechwan mountains, where they inhabit the moist gorges of small streams at 10,000–13,000 feet (3,000–4,000 m).

Here the ground is steep and precipitous, but there is still thick bush, forest or bamboo jungle. In Sumatra serow are found in the low tropical hills, always steep but well-watered and forested, where there are boulders and caves. Serow seem to migrate vertically with the season, going higher up the hills in summer. Japanese serow live at above 3,300 feet (1,000 m) in pine forest and open grassland, and the same is true for the Taiwanese serow.

Like the gorals, maned serows may move in small herds of up to seven animals, but they are more often solitary. Each serow appears to have a small territory; four or five will divide up a hillside between them. They are very active and, like many other mammals of the region, they feed mainly around dawn and dusk, lying up in cover during the heat of the day.

Goatlike courtship

In the Japanese serow, mating is preceded by a courtship ritual similar to that of both goats and gazelles. The male licks the female's mouth, strikes between her hind legs with his forelegs and rubs his horns against her genitalia. The season of birth varies in different parts of the range: in Japan births occur from June to August, in the Himalayas from May to June and in

The once endangered Japanese serow was down to 2,000 to 3,000 individuals, due to overhunting. Since it was declared a special national monument in 1955, its numbers have increased to 100,000.

The Japanese serow has a longer, thicker and more woolly coat than its mainland relatives and can tolerate lower temperatures and heavier snowfall.

Myanmar (Burma) at the end of September. Birth takes about half an hour, and the female walks about as the young is being born. The young is a foot (30 cm) high at birth and reaches full size in about a year.

Many predators

At the first sign of danger, serow dash away with a whistling snort. They are preyed upon especially by leopards and tigers, dholes (wild dogs), wolves and birds of prey, which take the young. In Japan and Taiwan most of these predators are absent, although there are still bears, and wolves have only recently become extinct in Japan.

Abominable Snowman?

In October 1953 four Indian mountaineers in the Himalayas visited the Nepalese monastery town of Pangboche, where they learned that the scalp of a Yeti, or Abominable Snowman, was kept in the local temple. They obtained permission to see and photograph this relic and to pluck a hair from it, which they sent for examination to Leon A. Hausman, an American zoologist. Hausman examined it and compared it with hairs of langur, bear and takin but could not identify it. Later that year Charles Stonor, formerly of the London Zoo, also examined the scalp. There was no trace of any sewing inside it, so in Stonor's opinion it was a scalp from a large crested, gorilla-like anthropoid. The hair was reddish with black bands and was bristly. A second scalp was said to be in the Khumjung monastery nearby, from a female Yeti, that in the Pangboche monastery being said to be from a male. In 1957 the Pangboche relic, in the care of a Sherpa, was allowed out of Nepal and brought to England. It was examined at the British Museum by a group of scientists. Alas for the hopes of the Yeti, the scalp was not sewn but molded—out of the skin of the shoulder of a serow, so we still do not know for certain if there is an unknown animal up there in the Himalayas.

SEROWS

CLASS **Mammalia**

ORDER **Artiodactyla**

FAMILY **Bovidae**

GENUS AND SPECIES **Japanese serow,** *Nemorhaedus crispus*; **Taiwanese serow,** *N. swinhoei*; **mainland serow,** *N. sumatraensis*

ALTERNATIVE NAMES
Kamoshika (*N. crispus*); maned serow (*N. sumatraensis*)

WEIGHT
110–310 lb. (50–140 kg)

LENGTH
Head and body: 55–71 in. (140–180 cm); tail: 3–6⅓ in. (8–16 cm)

DISTINCTIVE FEATURES
Stocky, generalized goat-antelope; short horns (both sexes)

DIET
Herbs; grasses; leaves, shoots and twigs of trees and shrubs. *N. crispus*: alder and sedge families; Japanese witch hazel; cedar.

BREEDING
Age at first breeding: 2–3½ years; breeding season: variable, according to region; number of young: 1; gestation period: about 210 days; breeding interval: 1 year

LIFE SPAN
Up to 20 years

HABITAT
Mountains with brush or thick forest

DISTRIBUTION
Japan, Taiwan and highlands of Southeast Asia from Nepal to Peninsular Malaysia and Sumatra

STATUS
Vulnerable or conservation dependent; some subspecies endangered

Serow ■ Mainland □ Taiwanese ■ Japanese

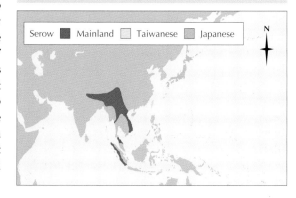

SERVAL

THE SERVAL IS ONE OF the most striking members of the cat family. Slenderly built, it stands 18–24 inches (45–60 cm) at the shoulder, with a combined head and body length of 25½–35 inches (65–90 cm) and a comparatively short tail of 10–14 inches (25–35 cm). The serval weighs about 19¾–39½ pounds (9–18 kg). It is long-legged and is noted for its large, erect, pointed ears. The smooth, short hair is yellowish brown on the back and sides and marked all over with a conspicuous pattern of bold black spots and stripes. The undersides are often white, the ears are black on the outside with a prominent white spot in the middle and the tail is ringed. As is common among cats, the upper part of the inner surface of each foreleg is marked with two black horizontal bands. Black melanistic forms of servals do occur, particularly near mountainous areas, although they are rare.

Although it was classified as a distinct species from the serval for some time, the servaline cat, *Felis servalina,* is now known to represent merely a second color phase of the serval. Its color pattern differs widely as the stripes and spots are replaced by a fine powdering of specks, giving a dusky or brownish hue to the whole body.

The serval is found in most of Africa, south of the Sahara. It prefers open country that receives relatively high levels of rainfall, often on the fringes of forests.

Swift and agile

The serval does most of its hunting by night and, although it may often be seen moving about in daylight hours, it usually prefers to curl up in a nest of grass during the day. Its long legs enable the serval to achieve great speed over short distances, and in high grass it progresses by long, high leaps. It is a skillful and rapid climber and can swim well. If attacked, it will fight vigorously and has occasionally killed dogs that have been hunting it.

Especially when they are young, servals make tame and gentle pets, playing in much the same way as domestic cats. However, adults of breeding age often become unpredictable, particularly in the presence of strangers. In common with domestic cats, tame servals purr when they are caressed.

Pounces on rodents

Servals live mostly on rodents. Like many other members of the cat family, they adopt the slink-run, ambushing and stalking type of hunting, which is particularly successful in catching small burrowing rodents. With such prey it is essential for the servals to remain hidden long enough for the prey to get well away from its burrow; otherwise it will go scuttling back to the safety of its hole. Servals conceal themselves in undergrowth and listen for sounds made by their rodent prey. They then leap high through the grass and pounce on their victim. Servals are highly efficient hunters, with a high ratio of kills to strikes.

In some areas servals catch and eat birds that they have flushed into the air or that are perched sometimes as high as 10 feet (3 m) from the ground. Being skillful climbers, they are able to capture and eat tree hyraxes. Servals also sometimes take lizards, snakes and insects. They frequently dispatch prey with a sudden downward, slapping blow of the outspread forepaw, delivered with considerable force and with the claws extended. Servals sometimes employ the same method when dealing with rodents.

Servals often hunt in marshy areas. It is thought that they may even help to keep down the numbers of coypu that have become established in dams in many parts of Kenya.

Relative to body size, the serval's legs are longer than those of any other cat. They provide the serval with elevation for hunting in tall grass and enable it to leap high and pounce on prey.

The serval's large, vertically set ears are highly efficient in tracking sounds, enabling the cat to accurately pinpoint the whereabouts of prey.

SERVAL

CLASS	**Mammalia**
ORDER	**Carnivora**
FAMILY	**Felidae**
GENUS AND SPECIES	***Leptailurus serval***

ALTERNATIVE NAME
Lynx tachete **(French);** *aner* **(Ethiopian);** *ingwenkala* **(Zulu)**

WEIGHT
19¾–39½ lb. (9–18 kg)

LENGTH
Head and body: 25½–35 in. (65–90 cm); shoulder height: 18–24 in. (45–60 cm); tail: 10–14 in. (25–35 cm)

DISTINCTIVE FEATURES
Pale yellow coat, marked with solid black spots; black bars on neck and shoulders; very long legs; large ears

DIET
Mainly rodents and other small mammals; also small invertebrates

BREEDING
Age at first breeding: 18–24 months; breeding season: year-round, peaks during wet season; number of young: 1 to 3; gestation period: 70–79 days; breeding interval: 1½–2 years, 1 year in wet areas

LIFE SPAN
Up to 19 years

HABITAT
Grassland, scrub savanna; absent from rain forest, but does occur in dry forest; also occurs in high-altitude alpine grassland

DISTRIBUTION
Widespread in sub-Saharan Africa, except where for a few rain forest localities

STATUS
Locally common

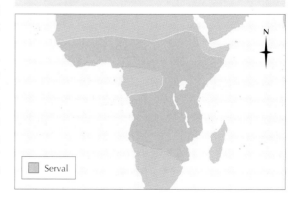

Serval

Servaline or serval?

The servaline was originally classified as a distinct species, and it was some years before scientists suspected that it might be only a color phase of the serval. In 1915 an African in Sierra Leone found two kittens, one a serval and the other a servaline, both said to be from the same litter. Some zoologists accepted this as evidence that the two represented one species. However, the late J. A. Allen, a leading American zoologist, argued that this was a dubious assumption. He pointed out that the collection of skins in the American Museum of Natural History could be sorted into two clear groups, one large-spotted (serval) and the other small-flecked (servaline) with no intergrading, suggesting that they must be two separate species.

Although Allen's argument was persuasive, it was eventually discredited. A Ugandan game warden carried out a survey of skins worn by Africans in Northern Rhodesia, and found much intergrading between serval skins. Some of these skins were sent to the Natural History Museum in London, where they were studied. They provided clear evidence that the servaline is merely a color phase of the serval.

SHAD

NATURALISTS BELIEVE THAT the shad must once have been a well-known fish; its English name dates from at least the 12th century. However, its numbers have suffered from the effects of modern civilization. There are two species living in European seas, the allis shad, *Alosa alosa*, and the twaite shad, *A. fallax*. Other species include the common, white or American shad, *A. sapidissima*, of Atlantic North America, the hilsa shad, *Tenualosa ilisha*, from India and Myanmar (Burma), and Reeve's shad, *T. reevesi*, from China, all of which are fished commercially. The alewife, *A. pseudoharenpus*, which derives its name from the Native American name, *aloofe*, is a related North American species.

Shad are herringlike fish with deep, laterally compressed bodies covered with small, silvery scales. The allis shad measures up to 2 feet (60 cm), the smaller twaite shad to 20 inches (50 cm). The hilsa shad also reaches 2 feet long, and the American shad grows up to 30 inches (75 cm) long. The upper jaw is notched in the midline with the lower jaw fitting into it, and the gill cover is marked by weakly ridged, radiating lines. There is a single dorsal fin, a low anal fin set well back near the tail, small pelvic and pectoral fins and a forked tail fin. Along the underside there is a row of keeled scales, almost like a row of spines. The twaite shad is the more common shad but was not recognized as a separate species until 1803 because it is so similar to the allis shad. The adult twaite shad often has a row of six to eight dark spots running along the flank from behind the gill cover; the allis shad may have a single spot. The only certain way to distinguish them is to count the gill rakers on the first gill arch. The allis shad has 85 to 130 and the twaite shad has 30 to 80.

Enter rivers to spawn

The allis shad is found from Norway through the Mediterranean to the Black Sea, the twaite shad from Iceland to the Mediterranean. The American shad, a native of the Atlantic seaboard, has been introduced three times into the Sacramento River in California and is now well established there. All shad enter rivers to spawn. The twaite shad has subspecies in the Rhône and the Italian Lakes Maggiore, Como, Lugano and Garda. There is a landlocked shad, the goureen, in the lakes of Killarney in Ireland. This is another subspecies of the twaite shad.

The allis and twaite shad have a deep blue back with silvery flanks. They have more yellow on the flanks than herring and are more solitary, shoaling mainly just for their spawning runs.

Some shad spawn immediately on entering fresh water; others swim far upstream to their favored spawning grounds. This picture shows an allis shad.

The American shad moves around in vast shoals of fish of about the same size, moving at the same speed and at the same distance apart, with almost military precision. The European shad may have shoaled in a similar fashion when they were more numerous.

Gill rakers filter food

Shad are plankton feeders, with sieves in their throats for sorting their food. The large number of gill rakers, particularly in the allis shad, filter out the very smallest plankton. Shad also take larger plankton such as copepods and crustacean larvae and also eat small fish. The allis shad, with its finer gill rakers, takes more crustacea, whereas the twaite shad, with fewer gill rakers, takes more fish. Shad feed mainly in the sea but also, to some extent, in the fresh or brackish water on the spawning migrations up rivers in spring. The allis shad goes farther up the rivers than the twaite shad and in this respect is more like the American shad. The twaite shad tends to spawn just above tidal limits, in brackish water, as in the Nile, where it is especially abundant.

The males arrive first on the spawning grounds. They mature earlier than the females, at two to three years instead of four to five in the females, so the spawning males are smaller. The eggs sink to the bottom and hatch four to eight days later. The fry grow rapidly and move downstream. The newly hatched allis shad, for example, are about ½ inch (1.25 cm) long. They grow to 5½ inches (14 cm) in a year.

Shad seem unable to surmount artificial obstacles such as weirs. Even the rafts of garbage that collect in some estuaries are a deterrent to their migrations upstream, while sewage and industrial pollution completely upset their spawning migrations. In British rivers such as the Severn and Irish rivers including the Shannon, shad are still caught with stake nets. However, they no longer come up the more polluted rivers such as the Thames.

Known by many names

The origin of the name shad, spelled "sceadd" in Old English, dates from Anglo-Saxon times. Today in England it is probably unknown to most people apart from those who study or catch fish. On the Continent the allis shad is often named after the month in which it migrates up rivers. In the Netherlands and Germany, where the migration takes place in May, it is known respectively as *Meivisch* and *Maifisch*. In North America, however, the name shad came to apply not only to the related American shad but also to related fish, such as the menhaden, genus Brevoortia, also called the green-tailed, hardhead or yellow-tailed shad, and to unrelated fish, such

as the gizzard shad, *Dorosoma*, and the skipjack shad, *A. chrysochloris*. In the United States, the name has also been used for flowers that bloom or animals that appear at the time the shad migrates up the rivers, such as the shadberry, the shadbush, the shad fly and the shad frog.

SHAD

CLASS	**Osteichthyes**
ORDER	**Clupeiformes**
FAMILY	**Clupeidae**
GENUS	**7 genera**
SPECIES	**31 species, including American shad, *Alosa sapidissima* (detailed below)**

ALTERNATIVE NAMES
Common shad; white shad

WEIGHT
Up to 11¼ lb. (5.1 kg)

LENGTH
Up to 30 in. (75 cm)

DISTINCTIVE FEATURES
Slightly prominent lower jaw; 59 to 73 gill rakers; dark spot on shoulder; dark metallic blue-green shading on back; silver below

DIET
Plankton; sometimes small fish

BREEDING
Breeding season: November–June; number of eggs: 100,000; spawns in fresh water; by autumn, young fish enter sea

LIFE SPAN
Up to 11 years

HABITAT
Mostly at sea; breeds in freshwater rivers

DISTRIBUTION
Much of North America

STATUS
Common

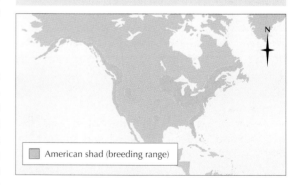

American shad (breeding range)

SHARKS

S HARKS ARE WITHOUT DOUBT THE best-known members of the cartilaginous fish, a class of fish that have skeletons made from flexible connective tissue rather than from solid bone as in bony fish. Other fish with cartilaginous skeletons include the eagle rays, devilfish, stingrays, guitarfish, skates, lampreys and hagfish. Since cartilaginous fish first appeared about 450 million years ago, they have adapted to every conceivable marine habitat. The bony fish are discussed in a separate guidepost article.

Classified by skeleton type

The cartilaginous fish (class Chondrichthyes) possess fewer varieties of body plan than the bony fish (class Osteichthyes). They are also less evolved than their bony relatives and consist of two main groups. The subclass Elasmobranchii contains the vast majority of cartilaginous fish species and accounts for 42 families of sharks, skates and rays, while the subclass Holocephali comprises just three families of chimaeras. The 12 families of skates and rays are definitely the most successful cartilaginous fish, with a diversity of about 450 species, representing about half of all living species in Chondrichthyes. There are about 350 species of sharks.

In addition to the two main groups of cartilaginous fish there is a separate class, Agnatha, which contains the two families of jawless fish. More primitive than either the bony fish or the true cartilaginous fish, the jawless fish possess some cartilage and a notochord (a tubular, longitudinal, flexible rod of cells) that provides most of their internal support. As their name suggests, the jawless fish have no jaws, but they also lack vertebrae, true fin rays, paired fins and even scales. All of

Most sharks live in coastal waters, but a few species, such as the blue shark (above), are found in the open ocean.

the jawless fish are highly specialized, leading unusual lifestyles. Most of the 45 species of lampreys (family Petromyzonidae) are parasitic on other cartilaginous fish, while the 43 species of hagfish (family Myxinidae) are equally parasitic, and also indulge in scavenging.

Different shapes

The shape of shark familiar to most people is the torpedo-like form of the swift-cruising requiem sharks (family Carcharhinidae), such as the Caribbean reef shark, *Carcharhinus perezi*, and the Zambezi or bull shark, *C. leucas*. Sharks lack the swim bladder found in bony fish and so body adaptations such as stiff pectoral fins have evolved to give them extra lift while swimming and to help them maintain a constant depth.

Bottom-dwelling species of sharks, rays and skates, together with the other species that lead sluggish lifestyles, have a totally different body plan. Collectively known as batoid elasmobranchs, the angelsharks or monkfish (family Squatinidae) and the skates and rays (order Rajiformes) possess

CLASSIFICATION
CLASS   Chondrichthyes
SUBCLASS   Holocephali: chimaeras;   Elasmobranchii: sharks,   skates and rays
NUMBER OF SPECIES   Approximately 1,000

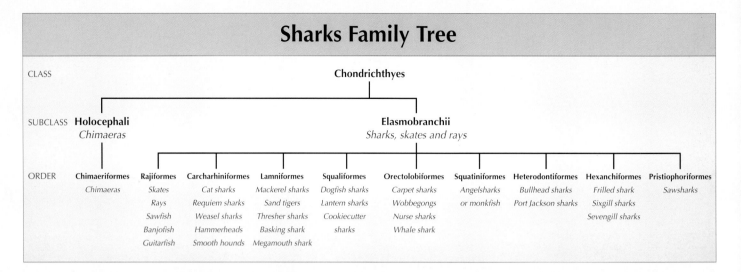

Sharks Family Tree

CLASS				Chondrichthyes					
SUBCLASS	**Holocephali** *Chimaeras*			**Elasmobranchii** *Sharks, skates and rays*					
ORDER	Chimaeriformes	Rajiformes	Carcharhiniformes	Lamniformes	Squaliformes	Orectolobiformes	Squatiniformes	Heterodontiformes	Hexanchiformes	Pristiophoriformes

	Chimaeriformes	Rajiformes	Carcharhiniformes	Lamniformes	Squaliformes	Orectolobiformes	Squatiniformes	Heterodontiformes	Hexanchiformes	Pristiophoriformes
	Chimaeras	*Skates*	*Cat sharks*	*Mackerel sharks*	*Dogfish sharks*	*Carpet sharks*	*Angelsharks*	*Bullhead sharks*	*Frilled shark*	*Sawsharks*
		Rays	*Requiem sharks*	*Sand tigers*	*Lantern sharks*	*Wobbegongs*	*or monkfish*	*Port Jackson sharks*	*Sixgill sharks*	
		Sawfish	*Weasel sharks*	*Thresher sharks*	*Cookiecutter sharks*	*Nurse sharks*			*Sevengill sharks*	
		Banjofish	*Hammerheads*	*Basking shark*		*Whale shark*				
		Guitarfish	*Smooth hounds*	*Megamouth shark*						

bodies that are almost completely flattened in the vertical plane. Instead of the short, stout pectoral fins of sharks, the pectoral fins of batoid elasmobranchs are attached at the back of the skull and are greatly enlarged to give these fish their distinctive body shape. The tails of most rays are reduced in size and are not used for swimming as is the case with sharks. Instead, locomotion is provided by the undulation of the tips of the pectoral fins. Rays are almost always bottom dwellers, inhabiting offshore environments and often extending into inland freshwater habitats.

The most notable deviation from the two main body plans just described occurs in the frilled shark, *Chlamydoselachus anguineus*. This species has an elongated body a little like that of an eel, and it lives in midwater.

The plow-nosed chimaera, Callorhynchus milii, *uses its unusual trunklike snout to probe the seabed for food.*

Body armor

The dermis, or skin, of elasmobranch species is extremely tough and flexible. Together with the cartilaginous skeleton, the skin serves an important role in maintaining the rigid body shape of these fish, particularly in the head region.

There are no true scales covering the thick dermis. Instead, hardened enamel "scales" known as denticles add a further layer of rigid armour. Packed closely together, the small denticles have a toothlike structure and, with their tips facing backward, are abrasive when rubbed against the direction of the water flow. In species of rays such as the thornback ray, *Raja clavata*, some of these denticles have become enlarged to form defensive thorns that protect the fish from predators. In the case of the stingrays (in the families Dasyatidae and Urolophidae) areas of denticles have become modified into venomous, daggerlike stings. The stings are used in defense rather than to stun or kill prey.

Coloration and camouflage

Compared to the rich range of coloration exhibited by bony fish, there is little variation in the skin tone of most elasmobranchs. With few exceptions, shades of brown, gray and black are dominant. Wobbegongs are among the more colorful species. These bottom-living sharks have varied coloration that conceals them against coral reefs, and species such as the tasseled wobbegong, *Eucrossorhinus dasypogon*, also have a variety of fleshy fronds to further disguise their outline.

Habitat requirements

Nearly all coastal waters of the world support at least some bottom-dwelling sharks, which play an important part in the ecosystems in which they live. Temperate regions are usually dominated by smaller requiem sharks and by stingrays, which frequent rocky reefs, muddy bays and open sandy habitats. Their major food sources are the vast abundance of bony fish and a selection of bottom-dwelling invertebrates. Tropical reefs are a favored haunt of species such as collared carpet sharks (family Parascylliidae), wobbegongs (family Orectolobidae), guitarfish (Rhinobatidae) and stingrays, although it is rare to find representatives of all of these families on a single reef. Unlike many species that are associated with a particular area, most of these sharks and rays are circumglobal in distribution and can be found over vast areas of ocean.

A few species inhabit the open ocean. The blue shark, *Prionace glauca*, is quite common in epipelagic (close to the surface) oceanic waters. Another visitor to the open ocean is the porbeagle shark, *Lamna nasus*, which often follows the migration patterns of its bony fish prey. In addition to the more aggressive meat eaters there are three families of very large sharks that feed on oceanic plankton. These filter-feeding

The stingrays (blue-spotted stingray, Taeniura lymna, *above) use their venomous daggerlike stings for defense.*

giants include the basking shark, *Cetorhinus maximus*, which grows up to 49 feet (12 m) long, and the whale shark, *Rhincodon typus*. The basking shark has minute teeth adapted to its plankton diet, huge 2-inch- (10-cm)-wide gill slits that almost encircle the body behind the head and unusual rakers on its hooplike gill arches, which filter plankton from the water. The whale shark is the world's largest fish, reaching up to 60 feet (20 m) in length.

Much remains to be discovered about the oceanic depths and the cartilaginous fish that live there. New species are constantly being discovered and classified. It was not until 1976 that the first known megamouth shark, *Megachasma pelagios*, was discovered off the coast of Hawaii, accidentally tangled in deep fishing nets.

Feeding behavior

Shark teeth are arranged in rows that are conveyed toward the front of the mouth on special folds of skin that form the gums. As the teeth become damaged or lost, new teeth continually move forward to replace them on the outer edge of the jaw. During their lives sharks can go through many thousands of teeth in this fashion.

The shape and structure of the teeth of cartilaginous fish offer an invaluable insight into their owners' diets. Many requiem sharks, such as the great white shark, *Carcharodon carcharias*, possess heavily serrated triangular teeth built for cutting flesh. The four species of sand tiger sharks (family Odontaspididae) use their special daggerlike teeth for piercing and tearing. By contrast, the bottom-dwelling cartilaginous fish, which feed mainly on hard-shelled crustaceans, have evolved teeth designed more for crushing and grinding than for cutting, piercing or tearing.

Some species of rays can discharge large quantities of electricity to stun their prey, although this ability is mainly used for defense. The pulses of electricity are generated by specialized tissue in the gill muscles located in the wings (modified pectoral fins). Most electric rays can electrocute a large fish with the pulses they discharge of 50 ampères at 50–60 volts. As with other electric fish, such as electric catfish and electric eels, the size of the voltage diminishes with repeated discharges. Larger electric ray species, such as the torpedo ray or Atlantic torpedo ray, *Torpedo nobiliana*, can produce a pulse of up to 200 volts. The jawless fish and a few species of sharks possess a poison apparatus in the form of modified mucus glands. While this adaptation is thought to be purely defensive in sharks, the hagfish use this mucus to incapacitate their victims before feeding.

Among the most unusual feeding behavior of the elasmobranchs is that displayed by the two species of semiparasitic cookiecutter sharks, genus *Isistius*. The cookiecutter sharks, seldom longer than 20 inches (50 cm) in length, prey on a wide range of larger cartilaginous fish and whales, attaching onto the skin of their victims with their suckerlike lips. They use their fused overlapping teeth to cut out a neat plug of flesh for them to feed on.

Some sharks are known to feed in groups according to an order of dominance established through body language. However, on occasions when many sharks are feeding on a single food source, such as a school of fish, a feeding frenzy may occur. Excited sharks dart in all directions, biting at whatever they can, including each other.

The sea lamprey, Petromyzon marinus. *Lampreys are primitive jawless fish that suck the blood of other fish. Like sharks, and unlike bony fish, they have a cartilaginous skeleton.*

Shark attacks on humans

Sharks often receive a bad press for attacking humans, but it is very rare for them to do this for food. Frequently the sharks in question will have been surprised when confronted at close quarters with a thrashing human. In such circumstances the sharks, especially those in the requiem family, usually exhibit a fierce threat display. If this threat display is ignored or unacknowledged by the human the shark may then be goaded into rushing its perceived threat with a fierce slashing attack in an effort to ward off the human. The three species most often involved in such attacks are the great white shark, the tiger shark (*Galeocerdo cuvier*) and the sand tiger *Carcharias taurus*. Others include the blacktip shark, *Carcharhinus limbatus*, bull shark and Caribbean reef shark.

The International Shark Attack File (ISAF) is a compilation of all known shark attacks. It is jointly administered by the Florida Museum of Natural History and the American Elasmobranch Society, a professional body of international workers studying sharks, skates and rays. The ISAF records that only 58 unprovoked shark attacks occurred worldwide in 1999. Unprovoked attacks are defined as incidents in which an attack on a live human by a shark occurs in the shark's natural habitat without human provocation of the shark. The yearly total of 58 unprovoked attacks is similar to the totals recorded in 1998 (54) and 1997 (60) and the decade's yearly average of 54. The meticulous records of the ISAF prove beyond doubt that sharks' reputation of being vicious man-eaters is almost entirely without foundation. The truth is that sharks almost never attack humans unless they are forced to.

Senses

Like bony fish, cartilaginous fish monitor their surroundings using a combination of touch, smell, taste, sight and hearing. The sense of smell is particularly well developed in most cartilaginous fish. Some sharks are able to detect minute traces of chemical substances in the water, sometimes at strengths as low as one part per million. Contrary to popular misconception, however, this highly developed sense is not balanced by a reduced quality of eyesight. Research has shown that elasmobranchs have well-developed eyes that give excellent vision. In fact vision is so important to some large oceanic sharks that they have evolved an extra nictitating fold, or eyelid. Just before the shark bites its prey this structure is moved across the eye to protect the eye beneath.

Cartilaginous fish also possess a lateral line system, just like bony fish, which picks up currents and disturbances in the water. The lateral line system comprises many sense organs, called

Wobbegongs spend most of their time lying on the seabed, where their superb camouflage makes them almost invisible to prey swimming overhead.

neuromasts, located within sensory cells projecting from the surface of the fish. In addition to the lateral line system, cartilaginous fish have evolved an electroreceptor system that detects the faint electric discharges generated by their prey.

Breeding

Mating in cartilaginous fish is not yet fully understood since observations of mating are rare, even in captivity. In the round ray, *Urolophus halleri*, mating occurs in a belly-to-belly position. In many larger shark species mating is preceded by close parallel swimming, with the males using their teeth to get a better grip on their mate for intercourse. It is thought that for this reason the skin of the female blue shark is four times thicker than that of the male. All male sharks and rays have a pair of elongated erectile clasper organs located on the inner part of the pelvic fins. One or both of these claspers may be inserted into the female's vent to transfer sperm.

Once fertilized, the female elasmobranchs carry their young in one of three ways. Approximately 40 percent of all cartilaginous fish, including many cat sharks (family Scyliorhinidae), all skates (Rajidae) and some rays, are oviparous; that is, they lay eggs externally inside leathery egg cases. Other species, such as the blue shark, are viviparous, with the yolk sac of each egg, once exhausted, becoming attached to the uterine wall in such a way that it can serve as a placenta. The babies are then born after hatching. In ovoviviparous species the thin-skinned eggs are retained inside the female until the young fish hatch; the babies are then born soon afterward.

After hatching, the young of some ovoviviparous sharks, such as the bigeye thresher shark, *Alopias superciliosus*, feed on any unfertilized eggs that they encounter before they emerge from the female. This is, in effect, cannibalism.

Cartilaginous fish take a number of years to reach maturity. During this time their main predators include larger specimens of their own species. Sharks, rays and skates never stop growing and many species are long-lived, with some living for several decades. The spiny dogfish, *Squalus acanthias*, has been known to live for at least 40 years. Lemon sharks, *Negaprion brevirostris*, can live for 50 years.

Conservation

By far the biggest threat to cartilaginous fish comes from humans. For instance, it is estimated that about 450,000 tons (408,250 tonnes) of sharks are caught by chance each year by boats fishing for other species. This is the equivalent of 26 million sharks. A further 75 million sharks are deliberately caught every year for food or for sport.

In many parts of the world shark fisheries are driven by the East Asian fashion for sharkfin soup. Many shark species are now threatened, and growing numbers are becoming critically endangered.

For particular species see:
- BASKING SHARK • CAT SHARK • DEVILFISH
- DOGFISH • EAGLE RAY • ELECTRIC RAY
- FRILLED SHARK • GREAT WHITE SHARK • GUITARFISH
- HAMMERHEAD SHARK • LAMPREY • MONKFISH
- PORBEAGLE • PORT JACKSON SHARK • SAWFISH
- SKATE • STINGRAY • THRESHER SHARK
- WHALE SHARK • WOBBEGONG • ZAMBEZI SHARK

SHEARWATER

Huge numbers of great shearwaters nest on islands of the Tristan da Cunha archipelago in the middle of the southern Atlantic. After breeding the birds migrate north to the North Atlantic.

THE 25 SPECIES OF shearwaters are petrels and belong to the family Procellariidae. They are relatives of the albatrosses and fulmars, with the typical tubenose nostrils and long, narrow wings. They are medium-sized seabirds, most measuring 12–18 inches (30–45 cm) long. The smallest species, Heinroth's shearwater, *Puffinus heinrothi*, is 9 inches (23 cm) long, and the largest, Cory's shearwater, *Calonectris diomedea*, is 21 inches (53 cm) long.

In most species of shearwaters the plumage is relatively dull, usually black or brown above and whitish underneath. The pattern and proportions of the colors, together with body size, are guides for the identification of shearwaters. The sooty shearwater, *Puffinus griseus*, for example, is very dark but with some grayish white underneath the wings, and the great shearwater, *P. gravis*, has a distinctive blackish cap and pale patches at the base of the tail.

Ocean wanderers

Shearwaters spend their lives at sea, being found in many parts of the world, but not in polar seas. Many shearwaters range widely, traveling regularly from one hemisphere to the other, whereas others have a fairly restricted distribution. The breeding grounds are on islands, where the shearwaters were free from predators until humans introduced new mammals, such as rats. Some breeding ranges are also restricted. For instance, the great shearwater breeds only on Gough Island and on the islands of the Tristan da Cunha group, with three to four million birds nesting on Nightingale Island. Heinroth's shearwater is known only from New Britain, the island to the east of New Guinea.

Effortless flight

Shearwaters are so named because of their flight, which is rather albatross-like. They skim low over the water following the contours of the waves, only rarely beating their wings. One great shearwater was seen to glide for 1½ miles (2.5 km) without a single wing beat. Exceptions to this flight pattern are the fluttering shearwater, *Puffinus gavia*, of Australia and New Zealand, and the little shearwater, *P. assimilis*, which has a broad distribution. Both of these species tend to beat their wings rapidly in quick bursts, interspersed with glides low over the waves.

During the breeding season shearwaters are not easily seen because they return to their nests in burrows at night. Some species, such as the

SOOTY SHEARWATER

CLASS	**Aves**
ORDER	**Procellariiformes**
FAMILY	**Procellariidae**
GENUS AND SPECIES	***Puffinus griseus***

ALTERNATIVE NAMES
Somber shearwater; ghostbird; New
Zealand muttonbird

WEIGHT
1½–2⅕ lb. (670–975 g)

LENGTH
Head to tail: 15¾–18 in. (40–45 cm);
wingspan: 37–42 in. (94–107 cm)

DISTINCTIVE FEATURES
Strong bill with hooked tip and large nostril
tubes; very long, narrow wings; webbed
feet; short tail; plumage almost totally
dark brown except for grayish white
patches on wing undersides

DIET
Mainly fish, squid and crustaceans

BREEDING
Age at first breeding: probably 5 or 6 years;
breeding season: eggs laid November–
December; number of eggs: 1; incubation
period: 52–55 days; fledging period: 65–75
days; breeding interval: 1 year

LIFE SPAN
Probably up to 35 years

HABITAT
Open seas and oceans; nests on islands

DISTRIBUTION
Breeds on subantarctic islands off New
Zealand, eastern Australia and southern
South America; spends southern winter in
seas and oceans of Northern Hemisphere

STATUS
Abundant

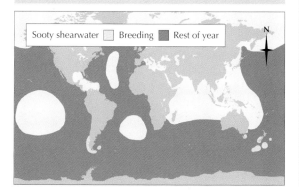

Manx shearwater, *P. mauretanicus*, even avoid visiting their burrows on moonlit nights. Outside the breeding season shearwaters can be seen far out to sea, although they rarely follow ships like albatrosses do, and it is often possible to see shearwaters from exposed cliff tops as they fly past on migration.

Migration

Many shearwaters perform very long migrations. The great shearwater flies northward from the islands in the middle of the South Atlantic in April to the coasts of Newfoundland and Nova Scotia, and then across the Atlantic to the coasts of western Europe, where it is seen mainly in August. A month or so later it returns the way it came. In the Pacific, the short-tailed shearwater, *P. tenuirostris*, undertakes a similar migration. From its breeding grounds in southern Australia and Tasmania it follows a figure-eight route, going first up the western side of the Pacific to eastern Siberia and then across to Alaska, and sometimes through the Bering Straits, down the western coast of America and across the Pacific to Australia. The journey covers some 20,000 miles (32,000 km) but is carried out so precisely that nearly all of the shearwaters arrive at their destination, the traditional breeding grounds, during the same 11 days every year.

Feed at the surface

Like other petrels, shearwaters feed on animals that live near the surface of the sea. Their prey includes a wide variety of fish, squid and cuttlefish, with smaller quantities of crabs and other crustaceans. Among the many types of fish

Shearwaters visit their nesting burrows at night to avoid predators such as gulls, skuas and jaegers. Pictured is an adult Manx shearwater.

caught are anchovies, sprats, herrings, pilchards or sardines, capelin and sand eels. Shearwaters take fry (young fish) as well as the adults.

Honeymoon flights

With the exception of the Christmas shearwater, *P. nativitatis*, of the Pacific Islands, shearwaters nest in burrows in large colonies. On Nightingale Island there are up to four million shearwaters nesting in just 4,000 acres (1,620 ha); the ground is so riddled with burrows that walking becomes difficult. The shearwaters may have to compete for nest space not only with their fellows but also with other burrowers such as rabbits, petrels and puffins. Courtship consists of a mutual rubbing of the bills and cackling, the latter producing a barrage of sound when thousands of shearwaters are calling together.

Just before egg laying most shearwaters disappear from their burrows on a pre-egg laying exodus, or honeymoon flight. For 2–3 weeks they go out to sea to feed and build up their food reserves so that the female can produce a single large white egg and the male can survive the first long incubation stint. The

egg is laid shortly after the female's return on a rough nest of coarse vegetation or pebbles. Incubation lasts about 52–55 days and each parent takes stints of 1–2 weeks.

At first the down-covered chick is brooded by day, but later it is visited only at night to be fed. Each parent visits it in turn and the chick pushes its bill into the parent's bill to receive food. The chick puts on weight rapidly and becomes very fat. The parents then desert the chick, which remains in the burrow living on its fat supplies until its feathers have grown. Manx shearwaters leave the nest when about 70 days old and great shearwaters when 80–100 days old.

The muttonbirds

The fat chicks of shearwaters have long been harvested in many parts of the world for their flesh and the oil obtained from their fat. The short-tailed shearwater is called the muttonbird in Australia, and the sooty shearwater is known as the New Zealand muttonbird, although the flesh of these species does not taste like mutton. A feature of shearwaters that assists the collectors is that the eggs are all laid at almost the same time.

When the breeding season is over, sooty shearwaters migrate nothward from their subantarctic breeding grounds, reaching the seas off Canada, Greenland and Japan.

SHEATHBILL

THE SHEATHBILL IS the only terrestrial shorebird that breeds on the Antarctic continent. That is to say, it is the only Antarctic breeder that has unwebbed feet and habitually lives on land, rather than coming ashore purely for the purpose of breeding. It is a pigeonlike bird, 16 inches (40 cm) long, with pure white plumage. The bill is conical with a greenish, horny sheath covering the base, for which the sheathbill is named. Around the base of the bill there is a group of pink fleshy knobs and there is a pink patch of bald skin under each eye. The legs are short but strongly built and each wing bears a short black spur.

The two species look very similar. The lesser sheathbill, *Chionis minor*, which has a black bill, breeds on subantarctic islands such as Kerguelen, Heard, Marion and the Crozets, which lie to the north of 60° S, the latitude that marks the Southern Hemisphere polar front, where cold polar waters meet and mix with subtropical waters. The American or snowy sheathbill, *C. alba*, breeds on islands of the Scotia Arc from South Georgia to the South Shetlands and on the Antarctic Peninsula south to the Palmer Archipelago, south of 60° S. Nonbreeding sheathbills occur in the Falkland Islands and along the coasts of South America as far north as the River Plate.

Reluctant fliers

Sheathbills are reluctant to fly, and when they are pursued they run rapidly on their stout legs, attempting to dodge their pursuers. However, they can fly strongly, with rapid wingbeats, like a pigeon. They usually fly fairly close to the ground, but regularly make flights across the Drake Passage from the Antarctic Peninsula and neighboring islands to South America.

The usual home of sheathbills is along the shoreline, but they desert this for penguin rookeries during the summer and also take to scavenging around Antarctic field stations for any edible rubbish. During the winter they live in small flocks. Sheathbills are very inquisitive and will follow humans or cats closely, but warily enough to scatter if they feel threatened. The birds quarrel frequently between themselves, often over morsels of food, and display to each other with rapid bows, at the same time uttering a low cackling or muttering sound.

Penguin parasite

Sheathbills feed mainly on animals and plants that live on the shore, although in the Antarctic winter this is covered with ice. On the shore they eat seaweeds, which they scrape off the rocks with their bills, and limpets, which they lever off. They also take small animals, dead fish and other carrion. When seals breed in early spring, sheathbills gather to eat the afterbirth, blood or any carcasses they find. However, the birds are best known for their habit of scavenging around penguin rookeries, where they run between the nests, dodging pecks from the penguins, in search of anything edible. In the summer the penguin rookeries provide a rich source of food, for sheathbills are omnivorous and exploit any available food source. Poorly guarded or abandoned eggs are broken open and dead chicks are torn apart. Sheathbills may kill very young or very weak chicks. They also eat penguin droppings and any krill spilled by the penguins when they feed their chicks.

Nests in crevices

Sheathbills return to their breeding grounds in October or November, and pair formation is accompanied by the same bowing and muttering displays that the birds use during quarrels or in the defense of their territories. The nest, a loose pile of penguin feathers, seaweed or limpet shells, is built under an overhanging rock or in a crevice near the shore. Both members of a pair

In addition to having a warming layer of fat about ⁴⁄₁₀ inch (1 cm) thick, snowy sheathbills ruffle up their feathers and stand on one leg to reduce heat loss.

Similar to a pigeon in shape, the sheathbill also has a pigeonlike flight, although it is primarily terrestrial. Its gait is similar to that of a rail, crow or gull.

assist in its construction, but the male is usually responsible for the collection of material, while the female arranges it. In early December she lays two, three or sometimes four eggs that are dirty white with dark blotches. The sheathbills share the incubation, which lasts for about 30 days. The chicks, which are covered with gray down, can walk fairly soon after hatching, but they stay in the nest for about 2 weeks. Later they wander about the territory. The parents guard them by attacking intruders, such as skuas, and the chicks hold still when their parents give the appropriate alarm call.

Stealing from babies

Sheathbill chicks are fed largely on krill. The ornithologist N.V. Jones, who made a study of sheathbills nesting around Adélie and chinstrap penguin rookeries, found that most sheathbills stole krill from the penguins by employing an ingenious trick. When a penguin is feeding its chick by regurgitating krill from its crop, the sheathbill flies up and flaps or pecks at them. The penguin chick jerks its head out of its parent's mouth to see what is going on, and as the latter cannot stop regurgitating immediately some krill is spilled. The sheathbill then lands and picks this up.

The Adélie penguins leave their rookery before the chinstraps, and their departure immediately cuts off a supply of krill. This could be disastrous to a sheathbill chick, but Jones found that one pair moved its feeding territory from an Adélie colony to a chinstrap colony when the former left. Sheathbills readily land on ships, particularly if there is food available. Ship-assisted birds have hitched rides in this manner as far as South Africa and even England.

SNOWY SHEATHBILL

CLASS	**Aves**
ORDER	**Charadriiformes**
FAMILY	**Chionididae**
GENUS AND SPECIES	***Chionis alba***

ALTERNATIVE NAMES
American sheathbill; pale-faced sheathbill; paddy

WEIGHT
1–1½ lb. (460–780 g)

LENGTH
Head to tail: 16 in. (40 cm)

DISTINCTIVE FEATURES
Plump, pigeon-shaped bird; all-white plumage; cone-shaped bill; dark gray legs

DIET
Seaweed, marine invertebrates, fish, eggs and chicks of other birds (especially penguins), carrion and scraps

BREEDING
Age at first breeding: 3–5 years; breeding season: eggs laid December–January; number of eggs: 2 to 4; incubation period: 30 days; fledging period: 50–60 days; breeding interval: 1 year

LIFE SPAN
Not known

HABITAT
Frequents shorelines, rubbish dumps, penguin rookeries and seal colonies; also follows ships

DISTRIBUTION
Breeding: Antarctic Peninsula; Scotia Arc (South Georgia to South Shetlands and Antarctic Peninsula, south to Palmer Archipelago). Winter: migrates north to Falklands and Tierra del Fuego.

STATUS
Locally common

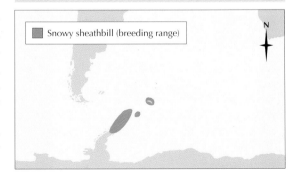

Snowy sheathbill (breeding range)

SHEETWEB SPIDER

SHEETWEB SPIDERS ARE known by a variety of names, including tunnel-web spiders and funnel-web spiders. These names derive from the fact that most sheetweb spiders build a small tunnel to one side of their web, in which they rest, awaiting their prey. Sheetweb spiders include the true house spiders, but not all sheetwebs are built in houses or even in buildings, many being found among foliage or herbage.

There are approximately 700 species of sheetweb spiders worldwide. They range in size from species that grow to a length of about 2 millimeters to specimens that have a leg span of 4 inches (10 cm). Sheetweb spiders use different methods to construct their webs, place them in a variety of locations and use them in different ways, according to species. However, in general, sheetweb spiders may be divided into those species that shelter in a tunnel web and run across the top of the horizontal web to seize their victims and those species that walk about upside down under their webs.

The most common species of sheetweb spiders in Europe are those of the genus *Tegenaria*. These are large, dark-colored house spiders with long legs. Of these, *T. domestica* was probably responsible for the original coinage of the word cobweb. It is brown, making it the palest in color of half a dozen species in the genus, and is now found all over the world.

Trip wire snares

As is common with small animals living in houses, sheetweb spiders originally made their webs in caves or similar enclosed places. *T. domestica* is found in houses and other buildings as well as in caves, hollow trees, between boulders or in any cavelike places. Among the handful of *Tegenaria* species, only experienced spider researchers are able to tell which species is responsible for a particular cobweb

The main difference between a sheetweb and an orb web (see orb spider, discussed elsewhere in this encyclopedia) is that the sheetweb is not sticky. Its upper surface has a tangle of trip wires spun above the surface of the web. In some species other strands are carried well up above the web. The result is that insects blundering into these fine silk threads are thrown down on to the web, where they have difficulty in regaining their balance because of the trip wires and are seized by the spider before they can escape. The sheetweb spider can run quickly over the web in spite of the trip wires, although how it achieves this remains a matter of scientific debate.

The funnel web in this photograph was made by the sheetweb spider **Agelena labyrinthica,** *which favors low vegetation and shrubs.*

2339

No escape for insects

Orb spiders respond to the vibrations of insects struggling in their webs. Sheetweb spiders quickly come out when the threads of the trip wires are pulled, an action typically brought about by an insect when it is striving to reach the edge of the web, for sheetweb spiders feed on crawling insects rather than flying insects. The spider kills its prey by biting and paralyzing it, then dragging it away to its tunnel.

Occasionally, the prey of the sheetweb spider tries to escape by biting its way through the web. However, this frequently results only in prompting the spider to run under the web and quickly seal the hole with more strands of silk before returning to claim its victim.

Courting by remote control

Broadly speaking, the courting patterns of sheetweb spiders are the same as those of other spider species. The male spins a tiny web on which he deposits his semen. He then takes this up in his palps, the pair of short appendages on the head. He goes in search of a female web and plucks it, in the same way as a struggling insect, gently dragging and pulling the trip wire threads of silk. Having done this, he withdraws his legs under his body, waits for a moment and then plucks again. If the female is receptive, she withdraws her legs in the same way. At this signal the male runs across the web to the female's tunnel, turns her on her side and inserts his palps, first one and then the other, into her

genital opening to impregnate her with his sperms. The female *T. domestica* lays her eggs in a dirty white silken cocoon suspended by a few silk threads from a ledge by her web. After mating the male and female may remain together until the male dies. Although this courting behavior is generally applicable to sheetweb spiders, the large number of species may result in some variation in the mating biology.

The sheetweb spider typically spins a flat web, which may be suspended from rocks, plants or other features of the terrain.

SHEETWEB SPIDERS

PHYLUM	**Arthropoda**
CLASS	**Arachnida**
ORDER	**Araneae**
FAMILY	**Agelenidae**
GENUS	**Many genera**
SPECIES	**Approximately 700 species**

ALTERNATIVE NAMES
Certain species: funnel-web spider, tunnel-web spider

LENGTH
Up to ⁹⁄₁₀ in. (2 cm)

DISTINCTIVE FEATURES
Hairy body; often has long legs; oval abdomen may bear dark spots, chevrons or bars; narrow cephalothorax (united head and thorax), with 8 eyes at front; typically makes flat web, with narrow tunnel at side

DIET
Arthropods, particularly insects and other arachnids

BREEDING
Courtship may involve male plucking female's web; round eggs, typically attached to web in egg sac; male and female may remain together after mating

LIFE SPAN
Varies according to species; some may survive for more than 1 year

HABITAT
Human habitations; varied natural habitats, including low-growing vegetation, decaying trees and grasslands; webs strung from variety of objects, including rocks and plants

DISTRIBUTION
Virtually worldwide

STATUS
Generally common

SHELDUCK

THE SHELDUCK IS A brightly colored, rather gooselike duck that is common in estuaries and on coasts. It is noticeably longer than the mallard, *Anas platyrhynchos*, being 23–26 inches (58–66 cm) in length, and is considerably bulkier. The male and female shelduck have almost identical plumages. At a distance they appear black and white, but closer examination shows the body to be mainly white with bold patterns of chestnut and black. There is black on the wings and the tip of the tail, and a broad chestnut band runs across the shoulders and around the breast and continues down the belly. The speculum, the band of color on the wings that is familiar as a bright blue patch on a mallard, is metallic green. The head and neck are also metallic green, and the bill, which in the male bears a prominent knob, is bright red.

The ruddy shelduck, *T. ferruginea*, an inland species, is quite different from the common shelduck. Its plumage is orangish brown with black on the wings and tail and white on the wing coverts. The bill and legs are black. The adult male has a black ring around the neck. Finally there is the paradise shelduck, *T. variegata*, also known as the paradise duck, painted duck or rangitata goose. In this species the female is brighter than the male, with a pure white head, chestnut body and mottled gray-brown wings. The male is much darker and drabber, with a blackish-green head.

Distribution

The shelduck breeds mainly in the northern half of Europe, including the British Isles, and east through Central Asia to eastern China. Smaller numbers breed in the Mediterranean region. The ruddy shelduck breeds mainly in Central Asia and China, and also in Asia Minor, North Africa and the extreme southeast of Europe. The paradise shelduck is confined to parts of New Zealand, especially on South Island.

Both the shelduck and the ruddy shelduck are gooselike in flight as well as in appearance, their flight being slower than is usual for ducks. The preferred habitats of the two species are very different, however. The ruddy shelduck lives inland and is only occasionally seen on the sea.

The common shelduck is a rather gooselike species of duck. It lives on coastal marshes and mudflats in Europe (above) and beside lakes in Asia.

Adult male shelducks have a prominent knob at the top of the bill.

SHELDUCK

CLASS	**Aves**
ORDER	**Anseriformes**
FAMILY	**Anatidae**
GENUS AND SPECIES	*Tadorna tadorna*

ALTERNATIVE NAME
Common shelduck

WEIGHT
Usually 2–3 lb. (905–1,350 g)

LENGTH
Head to tail: 23–26 in. (58–66 cm); wingspan: 3⅔–4¼ ft. (1.1–1.3 m)

DISTINCTIVE FEATURES
Bulky, gooselike body; elongated head; large red bill with (adult male only) prominent knob at top; mainly white plumage with dark green head, black patch on wings and broad orange-chestnut band across shoulders and around breast; pink legs; bright red webbed feet

DIET
Mainly mollusks, crustaceans and insects; also small fish, worms, algae and seeds

BREEDING
Age at first breeding: 3–5 years (male), 2 years (female); breeding season: early May–August; number of eggs: usually 8 to 10; incubation period: 29–31 days; fledging period: 45–50 days; breeding interval: 1 year

LIFE SPAN
Up to 15 years

HABITAT
Shallow coasts, estuaries, marshes and lakes

DISTRIBUTION
Coastal areas of northwestern Europe and Mediterranean, east across Central Asia

STATUS
Locally common

Its breeding grounds include the mountains of Asia, and it even breeds on the very dry slopes of the Altai Mountains in Central Asia. The shelduck, on the other hand, usually nests along the coast or, in Asia, beside large lakes.

Feasting on snails

The food of shelducks is mainly small animals that live on shores and mudflats, such as winkles, whelks, small crabs and sandhoppers. They also eat some worms, small fish, algae, grasses and other plants. Observations made on shelducks feeding around the coast of England showed that they eat large numbers of a small snail called the laver spire, *Hydrobia ulvae*. This species of snail is found in vast numbers on mudflats. At one place the incredible density of 645,000 snails per square foot (about 60,000 per sq m) was recorded. The snails move with the tide, floating in on the high tide and then crawling down as it recedes, feeding as they go. They then bury themselves until the tide rises again. The shelducks, together with groups of mallard and pintail, *Anas acuta*, feed on the snails as they crawl over the mud or probe for them if they are buried. The apparent dependence of shelducks on these snails leads to a high death rate in cold winters when the mudflats are frozen over and the snails are unobtainable.

The ruddy shelduck, meanwhile, eats mainly plants, and feeds on land more frequently than the common shelduck does. It supplements this diet with a variety of mollusks, brine shrimps, aquatic insects and locusts.

Nests in holes

Shelducks usually nest fairly near the sea. The nest of down and grasses is built in a hole in a ruined wall, hollow tree or haystack, or in the deserted burrow of a mammal such as a rabbit, fox, badger or marmot. The female makes the nest and incubates the 8 to 10 white eggs but the male stays nearby, defending the nest against intruders. When the female leaves the nest to make short trips for food, the male accompanies her. On her return she dashes to her nest, for with her conspicuous plumage she must avoid being seen by predators. There is often an escape tunnel or a hiding place near the nest where the female can retreat if danger threatens.

The eggs hatch in 45–50 days and, when dry, the chicks are led to water by both parents. The family usually stays together and may be joined by adults that have lost their own broods. At other times the broods join together in crèches.

Timing of the molt

When the chicks are independent, if not before, the adult shelducks molt their feathers. As with other ducks, all the flight feathers are shed at once and the shelducks become flightless. The shelducks, however, do not molt at the breeding ground but fly to a special molting ground. The molt takes about 6 weeks and after it is complete the shelducks fly home or, if home is in a cold part of Europe, they migrate south.

The ruddy shelduck lives on inland seas and lakes, usually far from the coast in desert, steppe and mountain country.

SHIELDBUG

Mating shieldbugs, Graphosoma lineatum. The bold colors of their wing covers tell other animals it is unwise to make a meal of them.

SHIELDBUGS ARE ALSO called stinkbugs, and with good reason. Their often bright colors warn predators of their potent and smelly chemical defenses. Shieldbugs represent a group of plant bugs comprising four families of the suborder Heteroptera. All are flattened in shape and some have an outline like that of an heraldic shield. Most are ¼–½ inch (6–13 mm) long, but the colorful red, black, orange and blue *Oncomeris flavicornis* of Australia, is 2 inches (50 mm) long. Shieldbugs are included in the great order of insects called the Hemiptera, which is the scientific name for bugs. All bugs are characterized by mouthparts formed for piercing and sucking, and are hemimetabolous, that is, they grow into adults by incomplete metamorphosis from nymphs, without a pupal stage.

Most shieldbugs have a superficial resemblance to beetles, but beetles have biting mouthparts, and they develop by complete metamorphosis, involving distinct larval and pupal stages. Shieldbugs also resemble beetles in using the hind wings for flying and the forewings as a protective covering for the hind wings. Not all shieldbugs can fly, and most of those that can fly, do so only in hot weather. In the shieldbugs each forewing is divided into two parts, a thick leathery basal part and a thin membranous area toward the tip. This results in the backs of these insects being broken up into patterns of triangles, which clearly distinguish them from beetles.

Shieldbugs are insects mainly of warm climates and are most numerous and diverse nearer the Tropics. For example, fewer than 40 species occur in Britain. Many more are found on the continent of Europe, especially toward the south, and the diversity increases greatly toward tropical Africa.

Useful and harmful selection

Almost all shieldbugs are found crawling about on the foliage of trees or bushes, or in low herbage, and many of them are found attached to particular species of plants on whose sap or fruit they feed. The birch, hawthorn and juniper shieldbugs take their names from their food plants, and the last two types feed mainly on the berries. Some are agricultural pests. One of the tortoise bugs, *Eurygaster integriceps*, is a serious pest of wheat in Russia, the Ukraine and western Asia. The green vegetable bug, *Nezara viridula*, has a worldwide distribution in the warmer countries, including southern Europe, and does great damage to beans, tomatoes and other vegetables. It is sometimes encountered in imported vegetables in northern European countries, but it does not establish populations in these countries.

In contrast to these harmful species, some of the shieldbugs are predatory and may be of service in destroying harmful insects. The North American genus *Podisus* is a useful predator of the Colorado beetle. The common northern European species *Picromerus bidens* feeds throughout its life on caterpillars but has no preference for any particular kind.

Broody shieldbugs

Shieldbugs lay their eggs either on their food plant or on the ground. The eggs look rather like those of butterflies and moths. They are usually developed in the insect's body, a few at a time, with the eventual total reaching 100 or more. A few shieldbugs develop their eggs in dozens and lay them in two neat rows of six. In many species the eggs hatch by the opening of a lid on the top, so that under a microscope the egg cases look like little empty barrels.

The young grow by stages, changing their skins usually five times before reaching full size. Although the development is gradual, there is often a startling change in color and pattern when the adult stage is reached. *Sehirus dubius*,

SHIELDBUGS

PHYLUM	**Arthropoda**
CLASS	**Insecta**
ORDER	**Hemiptera**
SUBORDER	**Heteroptera**
FAMILY	**Acanthosomidae, Cydnidae, Scutelleridae and Pentatomidae**
SPECIES	**Approximately 5,500**

ALTERNATIVE NAME
Stinkbug (applied to all species)

LENGTH
⅖–2 in. (1–5 cm)

DISTINCTIVE FEATURES
Shieldlike shape; flattened body; some species with striking colors; nymphs similar in appearance to adults

DIET
Plant sap and fruit; some species carnivorous

BREEDING
Hemimetabolous insects; nymphs molt several times (usually 5) before adulthood; number of eggs: up to 100 or more

LIFE SPAN
Weeks to years

HABITAT
The surfaces of plants

DISTRIBUTION
Worldwide except polar regions; most species in Tropics

STATUS
Some species abundant, many unknown

A clutch of young shieldbugs, such as these pentatomids in Sweden, may be actively protected by their mother before hatching, and even after hatching in some species.

quite a common shieldbug of continental Europe, is variable in color. In one form the young have bold black and red markings. After the last skin change, the adult is at first brilliant red, but after only a couple of hours its color darkens until it assumes its final livery of steely black.

A number of the shieldbugs are known to brood their eggs, attending and protecting them up to the time they hatch. The parent bug *Elasmucha grisea*, which lives in birch woods, goes further than this. The female lays a batch of about 40 eggs on a birch leaf, the egg mass being diamond-shaped and compact, the right shape for her to cover with her body. She broods the eggs rather as a hen does, for 2–3 weeks until the young hatch. The mother and small larvae stay around the empty egg shells for a few days while the mother actively protects her young, a behavior very unusual in insects. The family group then moves away in search of the birch catkins that form the main part of the diet.

Why stinkbug?

Many of the shieldbugs have glands from which they can eject an evil-smelling and ill-tasting fluid if molested. Anyone picking a berry and not noticing the shieldbug on it may get an unpleasant taste in the mouth. A bug held in one's fingers will usually resort to this same mode of defense. The smell is so strong and offensive that they are called stinkbugs, especially in North America. Some kinds feed on fruits and berries and render any they touch inedible to humans. The forest bug *Pentatoma rufipes* sometimes infests cherry orchards and spoils a great deal of the fruit in this way. It can be prevented from climbing up the trunks of the trees in spring by grease-banding the trunks.

Some species having this defense capacity are conspicuously colored, usually black with white, yellow or red patterns. They are undoubtedly examples of warning coloration. By making themselves conspicuous to predators, especially birds, they derive protection from the fact that a bird, once it has tasted one of the bugs, will remember its distinctive appearance and avoid trying to eat others of the same species.

Shieldbugs not protected in this way are preyed on by birds, especially tits (Paridae), which seek out the hibernating bugs in winter. Far more serious predators are the tachinid flies, whose larvae live as parasites inside the bodies of the developing bugs, killing them just before they reach maturity. These predators are in no way deterred by the bugs' repugnant fluids or lurid colors.

SHIP RAT

The ship rat is often found in wood-frame buildings, particularly in towns and cities in the Tropics.

THE SHIP RAT IS SOMETIMES referred to as the black rat, to distinguish it from the brown or common rat, *Rattus norvegicus* (discussed elsewhere in this encyclopedia). The ship rat is more slender than the common rat. It has a combined head-and-body length of 3¼–16 inches (8–40 cm), and the scaly-ringed and almost hairless tail is up to 10 inches (25 cm) long, longer than that of the common rat. The ship rat's weight is very variable and may reach up to 12¼ oz. (350 g). Its pointed snout projects far beyond the short lower jaw and its eyes are fairly large; its whiskers are long and black. The large, naked ears, contrasting with the finely haired ears of the common rat, are the best means of identifying the animal. The feet are pink, with scalelike rings on the undersides of the five toes and five pads on the sole. The thumbs of the forefeet are reduced to tubercles (prominent bumps).

In the past, the ship rat has been divided into three subspecies: *R. rattus rattus*, which is pure black above and black or dark gray beneath; *R. r. alexandrinus*, which is brown above and gray beneath; and *R. r. frugivorus*, which is brown above and white or cream beneath. It is more likely, however, that all are different forms of one species, and in Britain, where the ship rat is very rare, all three forms seem to live together and to interbreed so that many gradations of hair color exist.

Plague carrier

The ship rat probably originated in Southeast Asia and reached western Europe in ships returning from the Crusades in the Holy Land, about six centuries before the arrival of the common rat. Soon afterward, Europe was swept by the Black Death, a bubonic plague (pestilence caused by a bacterium and featuring swellings, or buboes). The term Black Death derives from the fact that the buboes on the victim, originating usually in the armpit or the groin, gradually blacken and spread over the body.

First recorded in Europe in 1347, the plague killed more than 25 million people in Europe and about 75 million people in the combined area of Europe, Asia and Africa. When the first wave of bubonic plague reached London, in 1348, it killed nine out of every ten inhabitants of the city. The plague is caused by the bacterium *Yersinia pestis*, which is carried by the fleas living in the ship

SHIP RAT

CLASS	**Mammalia**
ORDER	**Rodentia**
FAMILY	**Muridae**
GENUS AND SPECIES	***Rattus rattus***

ALTERNATIVE NAMES
Black rat; roof rat; tree rat

WEIGHT
1–12¼ oz. (28–350 g)

LENGTH
Head and body: 3¼–16 in. (8–40 cm)

DISTINCTIVE FEATURES
Black, brown, gray or olive upperparts; paler gray or white underparts; large, naked ears; long, virtually hairless tail of circular cross section; distinguished from brown rat (*R. norvegicus*) by more slender body, longer tail and naked ears

DIET
Virtually anything digestible; grain and root crops widely eaten

BREEDING
Age at first breeding: 80 days; breeding season: year-round; number of young: 4 to 11; gestation period: 20–24 days, more if female is already lactating; breeding interval: 3 to 7 litters per year

LIFE SPAN
Up to 4 years in captivity

HABITAT
Forests, farmland and urban environments, including docks and sewers

DISTRIBUTION
Worldwide except for cold, northerly latitudes, Sahara Desert and interior of South America and Australia

STATUS
Generally common; often a serious pest but becoming locally rare in some areas

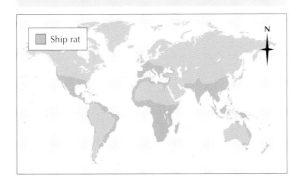

Ship rat

In common with other members of the genus Rattus, *the ship rat is capable of gnawing through almost any material, including thick rope and wood.*

rat's fur. Further outbreaks of the disease occurred during the following centuries, facilitated by the fact that ship rats were often found in ports and were regularly transported from country to country on ships. Even today bubonic plague is still a potentially fatal disease in some parts of the world, although it can now be treated with antibiotics and a vaccine.

The Great Plague of London (1664–1666), caused by the same disease, killed nearly 70,000 Londoners. The Great Fire that swept the English capital in 1666 may have contributed to a decline in the numbers of ship rats, since it destroyed many of the wooden buildings that they particularly favor and thus helped to rid the city of the plague. The spread of the common rat, which is heavier and more aggressive than the ship rat, probably also caused numbers to decline, and it is probable that this has occurred in agricultural habitats. The brown rat is also capable of carrying bubonic plague, but its burrowing lifestyle brings it into less frequent contact with humans and it is consequently less of a threat to human health.

The ship rat now ranges throughout the world but prefers the warmer climates, where it breeds rapidly. It is frequently found in urban

The ship rat favors warm places, such as the underside of roofs, barns and food stores. Because it destroys foodstuffs and transmits infectious diseases, it is regarded as a pest.

environments, usually in and around docks and among shipping crates. However, the strict de-ratting regulations now in force on ships are curtailing its traveling activities to a large extent. In Britain the ship rat is now confined almost exclusively to ports and to a few towns linked to seaports by canal. Being an expert climber, hence its alternative names of tree rat and roof rat, it is often found in the upper rooms of dock warehouses. The ship rat is mainly nocturnal but is also active by day.

In many places, the ship rat has been replaced by the common rat as a result of competition for food and living space. In rural areas in some parts of the Tropics the two species share the same habitat; the ship rat is the more successful in this location and has not been ousted by the common rat except, perhaps surprisingly, in some ports. The fact that wooden buildings are common in the Tropics may also explain the numbers of ship rats there.

Damage to stored grain
Like the common rat, the ship rat is omnivorous and eats anything that it can digest. It is particularly fond of fruits and grain, and in dock warehouses it does enormous damage to stored cereals, both by eating them and by contaminating them with its droppings. Although all rats

are associated with garbage, the ship rat is cleaner in its feeding than the common rat and maintains its own personal cleanliness by spending much of its time meticulously cleaning its fur and paws.

Large litters
Breeding occurs throughout the year, with a peak in summer and less often in the fall. The female collects quantities of suitable material such as rags, paper and straw and constructs a roomy nest, in which it rears three to five litters a year, often more. In the Mediterranean countries and India, where the ship rat lives a more outdoor life, the nest is usually made in trees. After a 21-day gestation period, the pink-skinned young are born, without fur, sight or hearing. They are usually born 7 at a time, but the precise number varies between 5 and 10. Ship rats become sexually mature at 3–4 months.

Effective de-ratting
In rural areas the ship rat, like the common rat, is preyed upon by larger animals. However, in urban environments its chief threat comes from humans. In recent years, rat destruction and prevention have become so effective that in highly developed countries, such as Britain, the ship rat is quickly disappearing.

SHIPWORM

THESE WORMLIKE BIVALVE mollusks, related to oysters, cockles and clams, have been regarded as pests by humans for centuries because of the damage they cause to wooden maritime constructions such as ships or piers. Shipworms may even have been instrumental in the failure of the Spanish Armada in 1588.

Many bivalve mollusks bore into rocks or into wood but none does so as efficiently as the shipworm. Shipworms range worldwide. Most measure about 10–12 inches (25–30 cm) but some are 3 feet (90 cm) long, especially those in tropical waters. Some individuals have occasionally been found measuring 6 feet (1.8 m) in length.

The two valves are very small, their surfaces are rasplike and they can be rotated to some extent. The body of the shipworm is long, slender and cylindrical, ending in a pair of short siphons that are supported by two calcareous (calcium-containing) plates called pallets. The shipworm burrows into wood, lining its burrow with a paper-thin layer of chalky material, as a sort of secondary shell.

Swarming larvae

Some species of female shipworms shed eggs into the sea; others retain them in a brood pouch in which the eggs hatch and from which the larvae are later freed. In either case, the eggs are fertilized by sperms shed by the males into the sea. The larva, or veliger, is spherical with a circle of cilia (hairlike projections) around the base and a short foot on one side. It soon grows a pair of valves, and a pair of siphons grows out on the opposite side to the foot.

The larva feeds first on the yolk within its own body. As this is used up, it settles on a wooden surface and starts boring. This may be within 36 hours of birth or later, according to species. In an experiment, scientists lowered some timber into the water of a harbor and discovered that in 5 months up to 40 larvae had settled on each square inch of wood. Some had bored to a depth of 4 inches (10 cm). The larva bores in with a pulsating, opening and closing shell movement, which makes a pinhole opening into which the larva creeps.

Imprisoned in its larder

Some scientists believe that many species of adult shipworms get their nutrition from detritus and plankton.

Young shipworms, however, start to feed on wood, which they rasp away in the process of making their tunnels. By rotating the valves on their forward ends, the shipworms make cylindrical tunnels they can never leave. The tunnels are driven deeper and deeper, increasing in diameter as the shipworms become larger, but the holes through which the shipworms entered remain the same size.

A shipworm tunnels steadily on, turning aside only when it is in danger of boring into the tunnel of another shipworm or when it meets a knot of wood. A piece of infested timber soon becomes a labyrinth of tunnels separated by very thin walls, and pressure on such timber can easily cause it to collapse.

Although the shipworm is imprisoned for the 7 or 8 months of its life span, it keeps in touch with the outside world by pushing its siphons out through the original entrance hole. One siphon brings in water with oxygen and living plankton, which supplements the diet of wood. The other siphon discharges the water laden with waste products, after it has circulated around the shipworm's body.

The shipworm can withdraw completely into its burrow and plug the entrance hole against predators with its pallets. Under these circumstances it can survive on food reserves in its body for several weeks.

Shipworms cause considerable damage to wood. Their tunnels may have small entrances but the burrows get bigger and bigger inside the wood as the shipworms grow.

Sometimes the shipworm Teredo navalis *(above) goes through several sexual phases, changing from male to female, back to male and perhaps even back to female again.*

SHIPWORMS

PHYLUM	**Mollusca**
CLASS	**Bivalvia**
ORDER	**Eulamellibranchia**
FAMILY	**Teredinidae**
GENUS AND SPECIES	*Teredo navalis*; **others**

LENGTH
Shell: up to ⅖ in. (1 cm); body: up to 12 in. (30 cm)

DISTINCTIVE FEATURES
Thin, fragile bivalve; animal rarely seen; borings 5–10 mm in diameter, lined with calcareous (calcium-containing) deposit

DIET
Fine, particulate material and plankton; probably little nutrition from wood

BREEDING
Sexual phases exhibited: first male, then female; eggs released into gill chamber; fertilized by sperm in water; brooded for some days; released larvae enter plankton before settling and developing on wood

LIFE SPAN
Up to 1 year, often much less

HABITAT
Typically drifted, waterlogged wood; also wooden piers, piles, groins and ships

DISTRIBUTION
Worldwide

STATUS
Abundant

Defeated by false bottoms

Shipworms attack any timbers immersed in seawater, such as pier piles, floating lumber or the hulls of ships and boats. The ancient Greeks tried to prevent the mollusks from causing damage by driving iron nails into a ship's hull below the waterline, overlapping the broad heads of the nails. The work would have had to be carefully done because recent experience in covering wooden piles or ships' hulls with copper or zinc sheathing shows that the larvae get in through any tiny gap between the sheets of metal. In the 17th century, ships' hulls were covered with a thin layer of expendable wood and the cavity between the false bottom and the true bottom was filled with a layer of cow's hair, which the shipworms could not penetrate. Later, a layer of felt was substituted for the cow's hair. However, in spite of antifouling paints, the use of steel ships and concrete pier piles, the problem of the shipworm has still not been solved.

Infested Armada?

Some historians believe that a major factor in the defeat of the Spanish Armada in 1588 may have been shipworms. A variety of causes have been suggested: superior English seamanship, faster ships and longer-ranged guns in the English ships, to name only three. However, the ships of the Armada were kept in Lisbon's saltwater harbor for months, which gave time for their timbers to become infested with shipworms. Then, as the fleet made its way north toward the English Channel, it had to sail at the speed of its slowest ships, which was less than 2 knots (3.7 km/h). Shipworm larvae, barnacles and other fouling organisms are able to settle at this speed. So the Armada may have arrived within sight of Plymouth, on the southwestern coast of England, already doomed from its crumbling underwater timbers caused by shipworms.

Shipworms of past eras

Fossil wood riddled with shipworms can often be found in the London clay, a sedimentary deposit around the Thames basin that was laid down about 50 million years ago. From examining these fossils there can be no doubt that shipworms similar to those living today were extremely numerous in the seas of that geological period. It is also reasonable to assume that shipworms were in existence much earlier than that. As ships and pier piles did not then exist, the shipworms could only have attacked tree trunks falling into the sea.

SHOEBILL STORK

THE SHOEBILL STORK is also known as the whale-headed stork; in Arabic, it is called *abu markub*, meaning "father of the shoe," referring to the bill, which resembles a clog. The shoebill resembles a heron, and its enormous, broad bill is even larger than that of the boatbill heron, *Cochlearius cochlearius* (described elsewhere). The concave upper mandible and convex lower mandible give the shoebill stork's head something of the shape of a blue whale, hence its alternative name. The shoebill stork stands over 3 feet (90 cm) high and has a wingspan of 8½ feet (2.5 m). The long, black legs bear long toes, and there is a short tufted crest, which is raised when the shoebill stork gets excited. The bill is olive green above and paler below, and it bears a hook on the upper mandible. The eyes are unusually large and have yellow irises. The plumage is slate gray with a greenish gloss.

The shoebill stork is found in parts of eastern central Africa, including Ethiopia, Uganda and Tanzania, but is not common anywhere in its range.

Heronlike life
Shoebill storks live in papyrus marshes and swamps where there are rafts of floating vegetation called *sudd*. They live singly or in small groups and are extremely heronlike in their habits. The birds stand motionless for hours on end, sometimes on one leg, with their heavy bill lowered against the neck.

When a shoebill stork walks, it does so with a slow, sure gait. If it is frightened, it takes off and flies low with slow wingbeats, its neck withdrawn so that the bill is resting on its breast. Shoebill storks usually settle again very rapidly after being disturbed and they seem to seek the safety of the papyrus thickets as quickly as possible. They can, however, soar strongly to a considerable height using their broad wings.

Lungfish diet
The shoebill stork searches for food in the swamps, standing still with its bill held close to the water or stalking very slowly. Prey is caught in the bill with a swift lunge and is hastily swallowed. Frogs and fish form the usual diet, and

the broad bill is used to dig lungfish out of the mud, being used as a scoop much in the same way as that of the boatbill. The shoebill stork preys on fish weighing up to 17½ ounces (500 g), the equivalent of its own body weight.

Breeding
Breeding is linked to water levels, and shoebill storks breed at the start of the dry season. They are solitary nesters, the nests being flat mounds of plant matter. Shoebill storks usually lay two eggs, although sometimes they may lay one or three eggs. Incubation lasts for about one month. The chicks do not have large bills. They fledge

The shoebill stork spends much of its time standing motionless in its native swamp habitat. However, it is a good flier and can soar for long periods on its broad wings.

SHOEBILL STORK

CLASS	**Aves**
ORDER	**Ciconiiformes**
FAMILY	**Balaenicipitidae**
GENUS AND SPECIES	***Balaeniceps rex***

ALTERNATIVE NAMES
Whale-headed stork; *abu markub* (Arabic)

WEIGHT
About 17½ oz. (500 g)

LENGTH
**Head to tail: about 4 ft. (1.2 m);
wingspan: about 8½ ft. (2.5 m)**

DISTINCTIVE FEATURES
**Massive, characteristic bill; long, dark legs
with long toes; slate-gray upperparts; paler
underparts; short tufted crest; large eyes
with yellow irises**

DIET
Mainly fish, especially lungfish; also frogs

BREEDING
**Age at first breeding: 3 years or older;
breeding season: varies with start of
dry season; number of eggs: usually 2;
incubation period: about 30 days; fledging
period: 95–105 days; breeding interval:
about 1 year**

LIFE SPAN
Up to 36 years in captivity

HABITAT
**Swamps and marshy lakesides, especially
with papyrus**

DISTRIBUTION
**Eastern central Africa, from southern
Sudan and Ethiopia to southern Zaire and
northern Zambia**

STATUS
Scarce

Shoebill storks are scarce and often difficult to observe in the wild. A captive individual is pictured.

after 95 to 105 days, but remain dependent on their parents for food for a week or more afterward. However it is rare for more than one chick to survive to fledging. The chicks are usually sexually mature after three years.

Classification problem

The first specimens of shoebill storks to reach Europe arrived in 1849. The bird posed a classification problem because it possessed the powder down patches of a heron but did not have the comblike claw on the third toe common to that bird. There have also been suggestions that it is a member of the Pelecaniformes, which includes the pelicans, gannets, cormorants and darters. DNA analysis indicates a close relationship with pelecaniform birds. At present scientists place the shoebill stork in a family of its own, Balaenicipididae, near the herons and storks.

Another puzzle set by the shoebill stork is its geographic distribution. Scientists have difficulty in establishing this because the shoebill stork is a shy bird living in swamps that to this day remain almost impenetrable, and it is extremely difficult to get close to one without alarming it into fleeing. The first specimens were obtained on the Bahr el Ghazal, a river in the Sudan; a little later scientists found shoebill storks on the White Nile. Other shoebill stork habitats have shown up only gradually, and the bird's known distribution is still patchy. The shoebill stork is considered to be near-threatened.

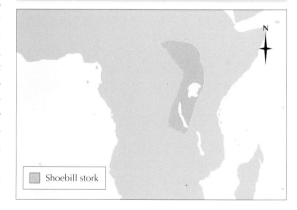

Shoebill stork

SHORE CRAB

THE CRAB BEST KNOWN on the western coasts of Europe is the shore crab, *Carcinus maenas*, a hardy, pugnacious animal the claws of which, upraised in self-defense, have earned it the name of *le crabe enragé* in French. It also lives on the shores of Atlantic North America, from Cape Cod to New Jersey, where it is known as the green crab. Shore crabs live from the highest rock pools to the zone of large brown seaweeds below low tide level, and have sometimes been found down to 650 feet (200 m). When fully grown, they vary from blackish green to reddish, lighter underneath than on top. However, when they are 1 inch (2.5 cm) across, the young crabs may sport patches of green, yellow, red or white in a variety of patterns.

In addition to the clawed legs, there are four pairs of walking legs without pincers. The back is covered by a single unjointed piece of armor called the shell, or carapace, the edges of which bend down and inward at an angle and are notched at the front. The abdomen, which is large in lobsters, is simply a flap tucked underneath the shore crab at the hind end. It has five joints in male shore crabs and seven joints in females.

At the front of the body are two stalked compound eyes and two pairs of antennae. Neither of the pairs of antennae is very long, but the first pair is particularly short. Around the mouth are six pairs of complex limbs, or mouthparts, each with a different role in feeding. Some of them are also used for respiratory purposes.

On the Pacific coast of North America the name shore crab refers to *Pachygrapsus crassipes*, along the coast of California, and to *Hemigrapsus oregonensis* farther north. Both are similar in appearance and habits to *Carcinus maenas*, but belong to another family, *Grapsidae*.

Working to an internal clock

The shore crab lives on rocky shores, sheltered mudflats and salt marsh pools, and is most numerous where shelter is available. It is the only European crab to thrive in estuaries and it can live in water with only one-sixth the salt concentration of seawater. It has a wide range of food, because the shore crab is a generalized predator as well as a scavenger and will eat anything it finds or can catch.

In summer, shore crabs tend to move up and down the shore with the rising tide, although sometimes they stay on shore as the tide ebbs. This behavior is particularly marked in the larger individuals. When shore crabs are removed from their normal habitat and kept in a laboratory in constant dim light, at a constant temperature, and either in moist air or continuously immersed, a complex rhythm of activity is observed despite the constant conditions of the laboratory environment.

Careful study has shown that this is due to the interaction of two components: a daily rhythm involving a peak of activity at night and a tidal rhythm in which the crabs are most active at the time of high tide, about every 12½ hours. One important consequence is that the crabs are most active at those times when they would be safest from drying up and from attacks by herring gulls. Scientists believe that both rhythms must be regulated by some kind of physiological clock in the nervous system.

When a crab is resting in a rock pool, a stream of water can be seen issuing from its front end, propelled through the gill chambers by paddles on the third pair of mouthparts.

With their flattened bodies, shore crabs can get into the narrowest of spaces. The shore crab in the picture above has taken over an empty scallop shell.

SHORE CRAB

PHYLUM	**Arthropoda**
CLASS	**Crustacea**
ORDER	**Decapoda**
FAMILY	**Portunidae**
GENUS AND SPECIES	***Carcinus maenas***

ALTERNATIVE NAME
Green crab

LENGTH
**Carapace (shell): 2⅛–2⅜ in. (5.5–6 cm) long,
3 in. (7.5 cm) wide**

DISTINCTIVE FEATURES
**Flattened, heavily calcified carapace,
ventrally fused in front of mouth; abdomen
small and reduced to a simple flap under
hind end of animal; 1 pair of massive claws;
4 pairs of walking legs; mature specimens
vary from blackish green to reddish**

DIET
**Plant and animal remains; also mollusks,
smaller crustaceans and fish**

BREEDING
**Separate sexes; eggs carried under female
abdomen; hatch into planktonic zoeae
larvae; undergo 6 molts; metamorphose into
megalopa larvae, then into small crabs**

LIFE SPAN
Up to 4 years

HABITAT
**Intertidal zone in rock pools, salt marshes
and estuaries; also shallow coastal waters,
occasionally down to 650 ft. (200 m)**

DISTRIBUTION
**Eastern coast of North America east to
Iceland, and western European coastline**

STATUS
Locally common

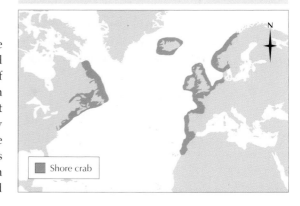

Shore crab

*Pregnant female shore
crabs carry their eggs
beneath their
abdomens. The eggs
hatch into free-
swimming larvae.*

The water enters through holes near the bases of
the legs. There are nine pairs of gills, springing
from the bases of the legs and some of the
mouthparts. The three hindmost pairs of mouthparts also bear comblike extensions for cleaning
the gills. Sometimes the direction of flow of the
water through the two gill chambers is reversed.
At other times, when the crab finds itself in a
pool depleted of oxygen, it can take in air
through the openings at the front, emitting
streams of bubbles farther back.

Crab metamorphosis

The eggs are carried under the female's flaplike
abdomen and pregnant females are found at all
times of year, although there is often a peak of
breeding that varies from place to place. In
southwestern England, for example, pregnant
female shore crabs are most common in February
and March. The larvae are quite unlike the
parents. They have well-developed abdomens
and each has a long spine at the front, between
the two big eyes, and another sticking upward

from its back. These planktonic larvae are called zoeae. Scientists originally believed that the larvae represented a species of crustacean that was different from the adult form. The zoeae molt about six times and then metamorphose into small crablike animals; this is called the megalopa stage. The megalopa swims with its third and fourth pairs of legs folded over its carapace, out of the way of its swimmerets, but it also can walk. Eventually it changes into a small crab, which goes on molting until it is about 3 inches (7.5 cm) across if it is a male, less if it is a female.

The shore crab's copulatory organs appear at the molt of puberty, when the crabs are about ⅔ inch (1.7 cm) across. After this there are 10 or 11 more molts, which occur with greater frequency in the warmer months of the year. The life span is about three or four years. John Vaughan Thompson, an army surgeon stationed at Cork, in southern Ireland, first traced the development of the shore crab in 1830. He also showed through similar careful studies that barnacles were crustaceans.

Consumed by humans

Shore crabs are preyed upon by bottom-feeding predatory fish, especially conger eels, and small crabs are often killed by the larger shore crabs. They also suffer significant losses between tidemarks, where they are attacked by gulls and by rats invading the shore at low tide. They are taken for human food to a minor extent in North America, as apparently they were in Europe at one time. Indeed, in 1895 there is an account of shore crabs being eaten by poorer people living around the Adriatic and also of being served as a delicacy at the tables of the rich. In the early 19th century, large numbers of shore crabs were sent to London, where they were eaten by the poor.

Full-time molting

In common with other crustaceans, a crab grows in stages, by molts. For a time it remains in a physically stable state, but then it sheds its outer covering, increases in size and then grows a new outer covering. However molting is not simply a brief interruption of normal life: the bodily changes associated with it continue during most of the crab's life. The shore crab's shell is made up of both calcium carbonate and organic matter. Some days before the molt, a large proportion of the shell's substance is reabsorbed into the body before the remaining part, now much more brittle, is cast off. During this time a new, soft shell is laid down

underneath. As the time of the molt approaches, the crab takes in water amounting to about 70 percent of its body weight. The crab's body swells and cracks the old shell along lines of weakness that are produced by the more complete reabsorption of tissue in some places than others. After the molt the crab, now soft-shelled, emerges and hides away until the shell hardens, because it is extremely vulnerable at this time. During this time, and for several hours afterward, it continues to take in water. There are also other processes of a more long-term nature. For example, between molts the crab continually lays down reserves of mineral salts in its liver. This is in preparation for the next molt.

X- and Y-organs

In recent years, zoologists have tried to find out how the crab's molting process is controlled. The shore crab, unlike the edible crab, does eventually cease molting, but it is possible to take a shore crab that has stopped molting and to make it molt again by removing its eyestalks. From many experiments scientists now know that the eyestalks produce, from an organ called the X-organ, a hormone that inhibits the activity of another pair of organs, called Y-organs. When uninhibited, the Y-organs produce a second hormone that triggers the sequence of events leading to molting. The X-organs are under neurological control and it is they that, in effect, stimulate the Y-organs to start the molt. At a certain age the spider crab, *Maia squinado*, also ceases to molt, but removing its eyestalks has no effect because the cessation of molting in this species is due to the degeneration of the Y-organ, and not to its sustained inhibition by the X-organs as is the case in the shore crab.

Shore crab larvae, also called zoeae, become part of the plankton before molting repeatedly and metamorphosing into small adult crabs.

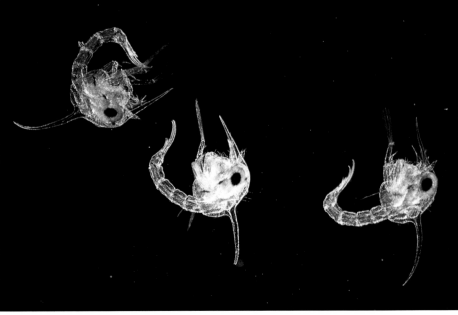

SHORT-EARED DOG

THE SOUTH AMERICAN short-eared dog is also sometimes called the zorro, after the Spanish word for fox. This alternative name has been applied to several of the South American dogs that are not strictly foxes but are foxlike members of the dog family. However, the short-eared dog does not appear to be closely related to any other member of the dog family. Nothing has been recorded of its behavior and habits in the wild and as yet only a few individuals have been kept in captivity.

The short-eared dog is so called because its ears are shorter than those of any other dog, apart from domestic breeds. They are 1¼–2 inches (3–5 cm) long and protrude above the crown of the head. The head and body measure 28¾–40 inches (72–100 cm), and the tail is 10–14 inches (25–35 cm) long. The head is large and the legs are short. Except for the maned wolf, *Chrysocyon brachyurus*, the short-eared dog is the largest of the South American dogs, standing about 14 inches at the shoulder. The hair is dark gray to black above and rufous mixed with gray and black beneath. The tail is long and bushy and is carried curled forward with the tip turned up to stop it sweeping the ground. The short-eared dog has long upper canine teeth that project ¼ inch (6.4 mm) when the mouth is closed.

Scientists believe that the range of the short-eared dog is limited to tropical South America. It is known to live in the Amazon basin in Brazil, Colombia, Ecuador and Peru, in the Orinoco basin in Colombia and probably Venezuela, and in the Rio Parana basin in Brazil.

Graceful catlike movement

The short-eared dog's existence was made known to science in 1882 when a live specimen arrived at London Zoo, and knowledge of its habits is virtually limited to observations that were made on a pair kept at Brookfield Zoo in Chicago. When a short-eared dog is excited, the hairs on the sensitive tip of its tail are raised, giving rise to another common name: flag-tailed dog. The male at Brookfield Zoo became extremely tame and would wag his tail feebly and roll over on his back, squealing, when petted by people he was familiar with. The female, by contrast, shunned human contact.

One of the most noticeable features of the short-eared dog's behavior is the catlike grace of its movements, quite unlike the plodding, stiff gait of many dogs. Scientists have yet to establish whether the short-eared dog is also catlike in its behavior, approaching its prey by stealth rather than by running it down, for instance. Its short

SHORT-EARED DOG

CLASS	**Mammalia**
ORDER	**Carnivora**
FAMILY	**Canidae**
GENUS AND SPECIES	***Atelocynus microtis***

ALTERNATIVE NAMES
Small-eared dog; short-eared zorro

WEIGHT
19¾–22 lb. (9–10 kg)

LENGTH
**Head and body: 28¾–40 in. (72–100 cm);
shoulder height: about 14 in. (35 cm);
tail: 10–14 in. (24–35 cm)**

DISTINCTIVE FEATURES
**Black or dark gray upperparts; reddish-black
and white underparts; thick, furry black tail;
short, rounded ears**

DIET
Some vegetation; small vertebrates

BREEDING
No details known

LIFE SPAN
Up to 11 years in captivity

HABITAT
Tropical forest up to 3,300 ft. (1,000 m)

DISTRIBUTION
**Amazon basin: northern Brazil, Ecuador,
Peru and Colombia; Orinoco basin: eastern
Colombia and probably southern Venezuela;
Rio Parana basin: southern Brazil**

STATUS
Rare

Short-eared dog

eared dog most closely resembles the bushdog, *Speothos venaticus* (discussed elsewhere in this encyclopedia), which scientists believe to be an excellent swimmer.

Both the short-eared dog and the bushdog have very small ears, which is surprising in view of the general tendency for an animal's extremities to become smaller toward the polar regions, as a comparison between the ears of the tropical fennec fox, *Fennecus zerda*, and the polar Arctic fox, *Alopex lagopus*, indicates. The reduction in ear size is related to the need to conserve heat by reducing the surface area of the body. It may be that the short-eared dog and the bushdog, both of which live in humid forests, have less need for large ears that facilitate efficient heat loss than the fennec fox of the deserts. Another theory is that large ears would form a conspicuous outline for prey species.

Unique among dogs

The male short-eared dog at Brookfield Zoo was often surrounded by a musky aroma from its anal glands, but there was hardly any noticeable aroma from the female. The male was one-third smaller than the female with a smaller head and a more slender muzzle, but he repeatedly dominated her, taking precedence at feeding times and not allowing her to feed until he was replete.

Apart from its superficial resemblance to the bushdog, the short-eared dog seems to be unrelated to any of the fox- or wolflike dogs. Its short ears, coat pattern, skull proportions and catlike gait, suggest that it is unique among dogs.

Like the short-eared dog, the crab-eating fox, Cerdocyon thous, is a foxlike dog native to northern and central South America.

legs are well adapted for easy locomotion in the dense forests of its native habitat. The coat is short and very sleek, and it is most likely that this is related either to a habitat in which there is frequent rain or to a partly aquatic lifestyle. Out of all the members of the dog family, the short-

SHREW

This pair of Eurasian common shrews may have a large meal of ant eggs in front of them, but shrews are so small that they cannot store large food reserves. Their metabolic rate is so high that in 2–3 hours they must eat again or face starvation.

THE SMALLEST SHREWS ARE the smallest terrestrial mammals. The only mammals that could be judged smaller than the smallest shrews are some tiny bats, like the bumblebee bat, *Craseonycteris thonglongyai*, of Thailand. The 320 or so species of shrews are classified into 22 genera, and are distributed across the world apart from in Australasia, the polar regions and much of South America. They bear similarities to the early insectivorous mammals that lived during the age of the dinosaurs. Shrews have been around for over 50 million years, and they have changed very little during this time.

Shrews of the genus *Suncus* are called white-toothed shrews, or musk shrews. White-toothed shrews include the musk shrew, *S. murinus*, which may be found in its own article. Another group of shrews is called the red-toothed shrews, and includes the common shrew of Europe and Asia, which is 4 inches (10 cm) long including a 1½-inch (3.8-cm) tail, and the European pygmy shrew, even smaller at 3¼ inches (9 cm) with a 1⅗-inch (4-cm) tail. The North American shrews include the common or masked shrew, *Sorex cinereus*, and the pygmy shrew, *S. hoyi*. These species are much the same size as their European counterparts. Several species of shrews are aquatic; they are described in a separate entry entitled "Water Shrew." Of the remaining North American shrews, mention can be made of the short-tailed shrew, which has a poisonous bite thanks to venom in its salivary glands; the smoky shrew, *S. fumeus*, whose tail swells in the breeding season; and the least, little or lesser short-tailed shrew, "the furry mite with a mighty fury" (E. Laurence Palmer).

The shrews described in this article do not vary much in appearance or habits. They are mouse-sized but have small ears and eyes, and a long, tapering snout bearing many long bristles. The fur tends to be grayish to dark brown on the back and dirty white on the belly.

Three-hourly rhythm

Shrews live solitary lives in tunnels in grass, among leaf litter or in surface tunnels. They are seldom seen and are only revealed by their high-pitched squeaks. There is reason to believe that shrews also use ultrasound clicks for echolocation but not to the high degree found in dolphins or bats. The common and pygmy shrews, and probably all the other species, have a 3-hourly rhythm of alternately feeding and resting. They continue this rhythm throughout the day and

SHREWS

CLASS	**Mammalia**
ORDER	**Insectivora**
FAMILY	**Soricidae**
GENUS	**22 genera**
SPECIES	**320 species, including Eurasian common shrew, *Sorex araneus*; European pygmy shrew, *S. minutus*; masked shrew, *S. cinereus*; North American pygmy shrew, *S. hoyi*; short-tailed shrew, *Blarina brevicauda*; and smoky shrew, *S. fumeus***

WEIGHT
¹⁄₁₄–1¼ oz. (2–35 g)

LENGTH
Head and body: 1⅓–7 in. (3.5–18 cm); tail: 3½–4¾ in. (9–12 cm)

DISTINCTIVE FEATURES
Small size; long, pointed snout; short, brown or gray fur; very small eyes

DIET
Mainly terrestrial invertebrates; some species: seeds and plant material

BREEDING
Age at first breeding: usually 1–2 years, sometimes within first year of life (female); breeding season: March–November in northern temperate regions; number of young: 2 to 10; gestation period: 17–28 days; breeding interval: breed 1 or more times a year, minimum interval 6 weeks

LIFE SPAN
1–2 years, perhaps longer in some species

HABITAT
Varied; usually moist microhabitats

DISTRIBUTION
Worldwide except polar regions, Australasia and much of South America

STATUS
Abundant to critically endangered

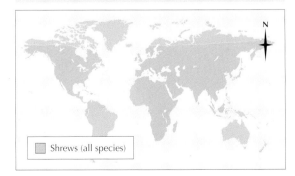

Shrews (all species)

night, but they are active for a longer proportion of each of the 3 hours during the night. Shrews are also relatively short-lived. In the common shrew and most red-toothed shrews, 15 months represents extreme old age.

There are many legends about shrews. One English legend is that a shrew cannot cross a human path and live, and some people have reported seeing a shrew tottering toward them and dropping at their feet. Such a shrew may have died of cold starvation.

Energetic knife edge

The smaller the animal, the greater its surface area in proportion to its bulk and the more readily it loses heat by radiation. This can only be replaced by food, so a very small animal must move restlessly, searching for food to make good the loss of heat—therefore the loss of energy—in searching for food. One result is that it cannot go without food for long periods; hence the 3-hourly rhythm. A shrew deprived of food for 2 or 3 hours will die, and the lower the temperature, the shorter the period of fasting it can endure. Most reports of shrews seen dropping dead relate to the early morning. This is the time, especially in autumn, when cold starvation is most likely to occur.

Death from shock?

The susceptibility to cold starvation led to the belief that shrews were especially prone to death from shock. Naturalists of the late nineteenth and early twentieth centuries reported shrews dying if a gun were fired near them or even if a blown-up paper bag were burst near them. In fact, when frightened, a shrew's heart can beat 1,200 times a

The European pygmy shrew is an animal that relies on the sense of touch. The eye is small and somewhat sunken, while the long, probing nose is covered in many long, sensory bristles. The nose can detect nematode worms ¹⁄₂₅–¹⁄₁₂ inch (1–2 mm) long.

minute, and they can die from the shock of loud noise such as thunder. When suffering from cold starvation, however, to the point where they are moribund, almost anything will kill them.

Bulldozers

Shrews are basically insect eaters but they will eat any animal small enough for them to overpower, such as snails and worms. They root out their prey from the soil, moss or leaf litter by using their long snouts as a kind of bulldozer. They will also eat carrion. In captivity, shrews have been found to remain healthy only if given a little cereal or seed each day, so they may also incorporate cereal into their natural diet. It is often said that shrews eat more than their own weight of food a day. A more exact estimate puts it at three-quarters their own weight.

High death rate

From spring to autumn each mature female has at least two litters of 2 to 10 young after a gestation of uncertain length that may be between 13 and 30 days. A newborn shrew weighs $\frac{1}{100}$ ounce (0.3 g). It is weaned shortly after its eyes open, at an age of around around 14–28 days. There is probably a heavy infant mortality in spite of the musk gland in each flank of most shrews, which emits a foul odor. Domestic cats, for example,

The short-tailed shrew of North America can easily subdue its prey thanks to the venom in its salivary glands.

will kill shrews but do not eat them, presumably finding them unpalatable. Birds of prey, especially various owl species, will eat them, however, and so will foxes and weasels.

Singing contests

There has long been the idea that shrews are not only ready to use their teeth when handled, but are also extremely quarrelsome among themselves. Naturalists have reported seeing two shrews apparently locked in mortal combat and squeaking furiously. Contests between European shrews have now been studied in detail. If two shrews meet they approach each other until their whiskers touch, then they squeak. As a rule, one of them, usually the intruder, will retreat. If it does not do so, both shrews rear up onto their haunches, still squeaking. If at this stage neither gives way, they throw themselves onto their backs squeaking even more and wriggling about. If the contest is still not resolved, the combatants often seize each other's tail in their teeth. The two continue wriggling and squirming, apparently in close embrace. Seldom do they hurt each other, or if they do, the injuries are not severe. Because food is so vital to them, they may become overcrowded and these singing contests are the best way to maintain a feeding territory under fierce competition.

SHRIKE

ALTHOUGH SHRIKES ARE members of the Passeriformes, or perching birds, they have some of the characters and habits of birds of prey. They are named after their shrieking calls. There is a rich variety of shrikes, but most are found only in Africa, south of the Sahara. Shrikes are divided into several groups. The bush-shrikes, which are confined to Africa, include the gonoleks; the puffbacks, with rumps like powder puffs; and the tchagras. The helmet-shrikes, also found only in Africa, bear a helmetlike crest and often nest socially. The true shrikes are at their most varied in sub-Saharan Africa as well, but one genus, *Lanius*, contains 13 species that range farther afield, including Europe and North America. True shrikes are also found in Asia as far south as New Guinea.

The *Lanius* shrikes are about 6–10 inches (10–25 cm) long with strong legs, sharp claws and a hooked bill. The bill has a tooth just behind the hook. The best-known shrike in North America is the northern shrike, *L. exubitor*. About 10 inches (25 cm) long, the northern shrike has distinctive plumage: steely blue-gray above, white underneath, with a broad black stripe running through the eyes and black-and-white wings. It has the widest distribution of all the shrikes, breeding in much of Europe, where it is called the great gray shrike, in Asia, and in Canada. Closely related, and similar in appearance, is the loggerhead shrike, *L. ludovicianus*, which is slightly smaller and breeds over much of North America from Manitoba to southern Mexico. The red-backed shrike, *L. collurio*, breeds through much of Europe and Asia from Britain to China. It is smaller yet, at 7 inches (18 cm) long, and the male has a bluish gray head and nape and a chestnut back. The female is duller.

Hunters

Shrikes are hunters. They swoop after small animals, strike them with their feet, and carry them away in their claws. They are such able hunters that they have sometimes been trained for hawking. King Louis XIII of France kept shrikes for hawking, and they have been used in the Near East and India. Northern shrikes have been trained in Alaska in order to study their methods of hunting. Their eyesight was found to be at least as acute as that of hawks. The shrikes could see bumblebees 100 yds (90 m) away, a distance at which they were invisible to humans.

The northern shrike has two main methods of hunting. It keeps watch from a perch and pounces on its prey, rather like a flycatcher. During its swoop down it may take advantage of cover to catch its prey unawares. It will also fly or run across the ground or through the undergrowth in an attempt to flush its prey. Shrikes occasionally chase their quarry in the air or will lie in wait for animals that have hidden.

Thorny larder

The shrike's best-known habit is that of impaling its prey on a thorn, barbed wire or other spiked object, or wedging it in a crevice. The habit is more common in the breeding season, when it

A northern shrike at work at its larder, a prickly bush where prey can be secured on thorns. Shrikes are not as strong as larger carnivorous birds, and this impaling behavior may help them break down their prey into manageable pieces.

An inhabitant of olive groves and cork oak forests around the Mediterranean, the woodchat shrike, Lanius senator, includes a range of small birds in its diet.

LOGGERHEAD SHRIKE

CLASS	**Aves**
ORDER	**Passeriformes**
FAMILY	**Laniidae**
GENUS AND SPECIES	***Lanius ludovicianus***

ALTERNATIVE NAME
Butcherbird (applied to shrikes in general)

WEIGHT
Average 1⅔ oz. (48 g)

LENGTH
Head to tail: about 9 in. (23 cm); wingspan: 12½–13 in. (32–33 cm)

DISTINCTIVE FEATURES
Black face mask; black wings with small white patches; black tail with white outer tail feathers; white or pale gray underparts

DIET
Insects and their larvae; small vertebrates

BREEDING
Age at first breeding: 2 years; breeding season: March–June; number of eggs: 4 or 5; incubation period: 15–17 days; fledging period: 16–18 days; breeding interval: 1 year

LIFE SPAN
Not known

HABITAT
Semidesert, savanna, woodland edge, orchards and other cultivated areas

DISTRIBUTION
Breeds from Alberta, Canada, in the north through much of U.S. to southern Mexico. Northern birds winter farther south.

STATUS
Fairly common in south of range; scarcer in the north

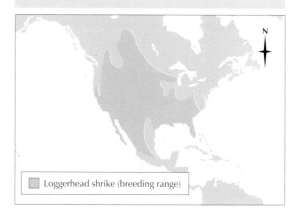

Loggerhead shrike (breeding range)

appears to be a method of storing surplus food for later use. The food is usually removed in a day or so. In New Mexico, however, loggerhead shrikes store lizards in times of plenty, and regularly return to them after several weeks have elapsed and cold weather has made live lizards hard to find. The conditions experienced by the northern shrike are colder yet, and the stored prey may be freeze while it is being stored. The northern shrike's ingenious solution is to tear up the prey shortly after killing it. The bite-sized pieces can be eaten even when frozen.

Butcher birds

Food is not always impaled; small animals may be held in the feet and torn up with the bill. However, prey is often impaled and eaten immediately when the shrike is hungry, in which case, the main function of impaling them is to secure the prey firmly so that the shrike can tear it apart. The shrike is much smaller than other carnivorous birds, and its feet alone may not be strong enough to hold the prey. The usual method of impaling is to hold the prey in the bill and pull it downward. Insects are impaled through the body and small mammals and birds through the chin. Impaling and wedging behavior is instinctive, as is demonstrated by hand-reared shrikes, but the shrikes have to learn suitable sites by trial and error.

The behavior of impaling prey and tearing it into pieces has earned shrikes the name butcherbird. However, they should not be confused with the Australian butcherbird, a member of the family Cracticidae, although the Australian butcherbird also impales its food on thorns or wedges it in clefts.

Varied prey

Shrikes feed on a large number of small animals. In at least some parts of their extensive range, northern shrikes concentrate on small birds, voles and insects. Small songbirds are caught frequently, including buntings, pipits, sparrows, martins and tits. Even young partridges may be killed. Most of the mammals killed are voles, but pipistrelle bats and young weasels have been recorded. Insects taken include bees, wasps and beetles.

Breeding

Nests are built in bushes, hedges or brambles. They are made of sticks, grass and roots with a fairly small hollow in the middle. In some species the nest is woven into a compact although bulky mass, and in others it is loosely woven. The male woodchat shrike courts the female by rapidly nodding his head and singing. The female joins in, and the pair form a duet. As in other true shrikes, the male continues his courtship by bringing the female food. If all goes well, the male will assist by feeding the female while she lays and incubates the eggs.

Efficient killers

Shrikes have a bad reputation because it is often thought that their larders are full of animals that they have killed needlessly, "for fun." In fact, nothing is further from the truth, for the shrike's usual way of killing its prey is by biting the neck, killing it almost at once. This method is very much like that of a falcon, and both shrikes and falcons bear teeth on the cutting edges of the upper mandible just behind the hook. The teeth may be used to penetrate between the vertebrae of the prey and damage the spinal cord, so enabling a falcon or shrike to kill larger prey.

A male red-backed shrike returns to his vantage point after a successful hunting sortie. The barbed wire would make a suitable larder.

SHRIMP

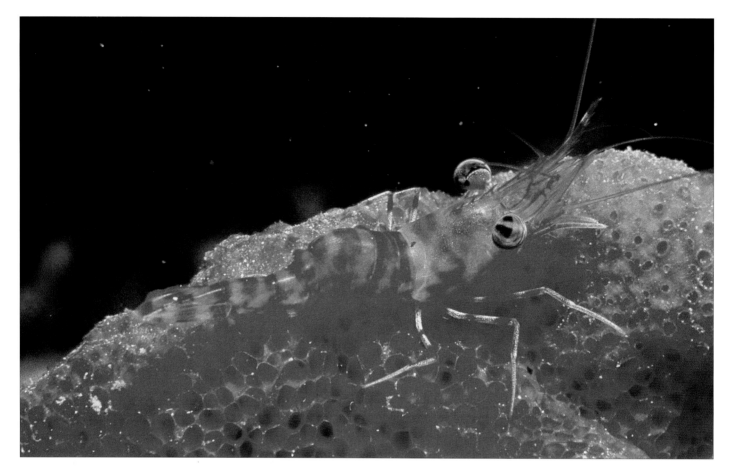

A red shrimp, Peneus duorarum, *on a red sponge. Shrimps can live in both fully marine environments and in fresh water.*

SHRIMPS ARE SMALL crustaceans that live on sandy bottoms in shallow seas off the coasts of Europe. The name has been given to many other crustaceans of similar size and appearance in different parts of the world, including some living in fresh waters.

The common European shrimp, *Crangon crangon*, is the species to which the name shrimp was first applied. It is up to 3½ inches (9 cm) long, and is gray to dark brown with dark brown or reddish dots. It lacks the saw-edged rostrum or beak in front of the head that characterizes the prawn, *Palaemon serratus* (discussed elsewhere), with which it is easily confused. The shrimp's outer pair of antennae are as long as its body and grow from a pair of oval plates on the front of the head. The inner pair of antennae are short and branched. The first pair of legs are short and stout and instead of pincers each leg has a movable spine that curves over to perform the same function as a pincer.

This same species also occurs on the Atlantic coast of North America, where it has been given the scientific name *Crago septemspinosus*. Both this species and the shrimp *Peneus setiferus* of the Gulf Coast, are fished commercially; 100,000 tons

(90,000 tonnes) of the latter are caught each year. On the Pacific coast the shrimp *Crago franciscorum* is fished commercially.

There are other crustaceans called shrimps that are really only shrimplike organisms. The European freshwater shrimp *Gammarus* is only very distantly related to the true shrimp; many marine species are related to the *Gammarus*. The giant freshwater shrimp of the United States, *Macrobrachium carcinus*, weighs up to 3 pounds (1.4 kg), in marked contrast to its common name: shrimp means small or shriveled. Cleaner shrimps, family *Hippolytidae*, remove parasites from some fish species, including the barber fish. Other shrimplike organisms include the sewing shrimps and the pistol shrimps, genus *Alpheus*, which stun prey with the sound waves created when they snap their large claws together.

Hardy sand dwellers

By day, or at low tide if it lives on the shore, the common shrimp buries itself in sand or mud, unless the water is turbid (murky). It shuffles with its walking legs while the swimmerets under its tail beat rapidly, driving the sand backward. At the same time it forces water sideways out of

EUROPEAN SHRIMP

PHYLUM	**Arthropoda**
CLASS	**Crustacea**
ORDER	**Decapoda**
FAMILY	**Crangonidea**
GENUS AND SPECIES	***Crangon crangon***

ALTERNATIVE NAMES
Brown shrimp; edible shrimp

LENGTH
Up to 3½ in. (9 cm)

DISTINCTIVE FEATURES
Lacks a rostrum, therefore has distinctive appearance compared with most prawns; gray-brown in color with brown spots; 2 pairs of antennae, outer pair as long as body

DIET
Mainly small animals and detritus; some seaweeds, particularly green seaweeds

BREEDING
Male when young; becomes female after mating; breeding season: spring and summer; internal fertilization; number of eggs: up to 14,000; fertilized eggs carried in appendages beneath abdomen; spawned some time after mating; complex series of larval and juvenile stages

LIFE SPAN
Up to 5 years

HABITAT
Sand from average tide level to around 164 ft. (50 m) deep

DISTRIBUTION
Coastlines of Europe, Mediterranean and North Africa

STATUS
Locally common

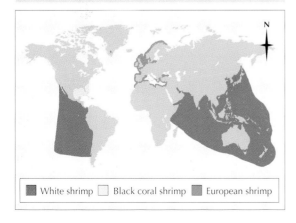

■ White shrimp　□ Black coral shrimp　■ European shrimp

its gill chambers and sinks into a trough. The long outer antennae are then swept around and backward, throwing sand over the body until the shrimp is completely hidden except for its short inner antennae, which stick up out of the sand. After dark the shrimp comes out, normally walking about on the last two pairs of thoracic legs. The other three pairs of legs are tucked in along the body, although the third and longest pair may be used as organs of touch. The shrimp may also swim just off the bottom, paddling along with its swimmerets or darting backward by suddenly bending its tail forward.

The common shrimp can tolerate a wide range of temperatures and salinity. It can live in freezing water, and tolerate water up to 87° F (30° C). It is found from fully marine coasts to salty estuaries and even in fresh water. However, it cannot survive the low winter temperature of brackish or nearly fresh water, so it migrates to the sea, the males being the first to go.

Shrimps eat mainly animal food but they also are known to feed on the smaller seaweeds, especially green ones, their diet varying during the year. They also take small crustaceans and mollusks, fish eggs and larvae as well as ragworms, which are found in large numbers in mud or sand.

Five larval stages
Mating begins when the shrimps are a year old and, except in estuaries where there is only one brood a year, each female has two summer

Periclemenes amethysteus on an anemone. This shrimp usually lives on two species of anemones, Cribrinopsis crassa and Anemonia viridis.

broods and another in winter. The winter eggs, 0.4 mm in diameter, are slightly larger than the summer eggs and take longer to hatch. As the eggs are laid the tail is bent forward and they are glued in a mass to the swimming legs. Small females lay 1,500 to 6,000 eggs, the larger females 9,000 to 14,000. The zoeae larvae, on hatching, swim to the surface and go through five stages before changing to young shrimps, which then sink to the seabed.

As well as being fished by humans, shrimps are preyed on by many marine animals, such as cuttlefish and fish. Shrimps are also often left behind in rock pools by the ebbing tide and are eaten by shorebirds.

Sewing shrimps and pistol shrimps

Shrimps bury themselves in the sand or burrow. However, one, *Alpheus pachychirus*, makes branching tubes 12 inches (30 cm) long and ¾ inch (19 mm) in diameter in tangles of seaweeds, each tube having a rounded chamber at the blind end. To make a tube the shrimp sews the edges of the weed together using very slender algae as thread. The shrimp lies on its back in a fold or furrow of the interlaced seaweed and, using its slender second pair of legs, pulls the two edges together. Then it thrusts one of these legs

through one edge, like a needle, catches hold of the algal thread from the opposite edge and pulls it. These first stitches are made at intervals, to hold the edges of the weed together; the shrimp then goes back and stitches between them. It can sew about 4 inches (10 cm) of tube in 10 minutes.

Pistol shrimps grow to 1–2 inches (2.5–5 cm) long and live in burrows in the sand. They have one very large claw, which may be either on the left-hand or the right-hand side and is more than half the size of the body. Instead of the usual pincers at the end of the limb, the claw features a structure that looks very much like the hammer and powder pan of an old-fashioned pistol, and is used in much the same way. The hammer is held at right angles to the limb and may be snapped back with force onto the pan. It is used against predators and to get food. Even a moderately large fish, 6 inches (15 cm) or so long, passing the mouth of a pistol shrimp's burrow, may be stunned by the sound of a pistol shrimp's claws snapping together, then seized and eaten. In experiments, scientists have discovered that a pistol shrimp snapping its claw in an aquarium will make a sound audible at the other end of a large laboratory. A pistol shrimp in a large glass jar has been known to break the jar with the sound waves created by its snapping claw.

An emperor shrimp, Periclemenes imperator, hitches a lift on a nudibranch, a type of marine sea snail.

SHRIMPFISH

THE SHRIMPFISH IS A relative of the rather curious sea horses and seems to be equally remarkable. To begin with, shrimpfish are so thin that when they turn sideways among weeds they seem to disappear. The whole head and body of a shrimpfish is flattened laterally (from side to side) and is completely encased in a transparent armor made up of a number of thin bony plates, fused with the underlying ribs in much the same way as the shell of a tortoise. On its underside these form a knife edge; hence the name razorfish, frequently used in Australia. At the hind end is a long spine. The two dorsal fins, which normally lie along the back of a fish, are crowded together beside the spine, the second fin actually pointing downward. The tail has also been pushed out of the position found in less specialized fish. It lies at an obtuse angle to the trunk and ends in a small tail fin also pointing downward. The snout is long and tubelike with a tiny, toothless mouth at its tip.

There are two genera of shrimpfish and two species in each genus. In the genus *Centriscus* the dorsal spine is long and solid, but in *Aeoliscus* it is jointed and movable. *A. strigatus* has been observed resting at the surface, head downward, with the tip of the spine turned at right angles to the body, which possibly gives it some stability through contact with the surface of the water. The spine contains nerves and sensory cells, so researchers suspect the spine also has some kind of sensory function.

Most shrimpfish are small. The largest is the Australian smooth razorfish, *Centriscus cristatus*, which can reach 12 inches (30 cm) in length. The body is silvery with a deep red band running from the mouth to the eye and continuing from there as an orange line along the flank to the dorsal spine. The belly is pale yellow with around 12 oblique red bars on it. Shrimpfish are found in shallow waters from East Africa to Hawaii. They are completely absent from the Atlantic Ocean.

Upside-down swimmer

Like the seahorse, shrimpfish orient their bodies vertically and move by slight movements of their fins. They rarely swim horizontally. With the dorsal and anal fins bunched together, shrimpfish are best adapted for swimming with the head down and the back pointing forward. This arrangement might seem awkward, but in fact, shrimpfish are rather agile, darting about with great rapidity. They can orient themselves both head-downward and head-upward, but they usually feed head-downward. A shoal of shrimpfish have been observed to feed while swimming with heads down along the sea bed. When they reach a vertical surface of rock or coral they continue in a horizontal orientation, heads remaining pointed towards the surface, even if the vertical surface becomes an overhang, in which case the fish are oriented with their heads upward. Perhaps their food is concentrated within 1–2 inches (2–5 cm) of rocks and coral, where water currents are slower.

Aeoliscus strigatus, a species of shrimpfish known as the razorfish, forms a typical small school among vegetation. Swimming head-down is the most natural way for shrimpfish.

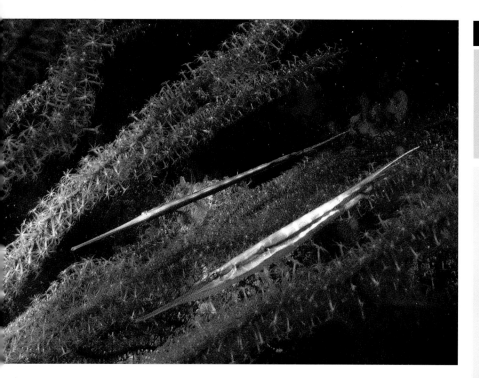

Although most at home around sea urchins, any linear feature of a shrimpfish's environment seems suitable to increase their concealment. Here, razorfish make use of gorgonian coral.

CLASS	**Osteichthyes**
ORDER	**Sygnathiformes**
FAMILY	**Centriscidae**
GENUS AND SPECIES	***Aeoliscus strigatus***

ALTERNATIVE NAMES
A. strigatus: razorfish; *A. punctulatus*: speckled shrimpfish; *C. cristatus*: smooth razorfish; *C. scutatus*: grooved razorfish

LENGTH
Up to 6 in. (15 cm)

DISTINCTIVE FEATURES
Body extremely flattened laterally and encased in armor of thin, transparent plates; mouth tubular and toothless; coloring yellowish with black band from snout through eye to base of tail; vertical, upside-down swimming behavior

DIET
Tiny planktonic crustaceans

BREEDING
Poorly known

LIFE SPAN
Not known

HABITAT
Marine: shallow coastal waters; among the spines of sea urchins, genus *Diadema*, or staghorn corals

DISTRIBUTION
Shallow waters of Indian and western Pacific Oceans: Sumatra in the west to Palau and Pohnpei (Micronesia) in the east; New South Wales and New Caledonia in the south to Korea and Japan in the north

STATUS
Uncommon to locally common

Disappearing tricks

When threatened, shrimpfish can streak away, swimming in a horizontal position like most fish. However, their first reaction to danger is to turn the sharp edge of the body toward the intruder and virtually disappear from its sight. Another common protective trick is for several shrimpfish to hang downward among the long slender spines of sea urchins of the genus *Diadema*. The spines are about as thick as the reddish line along the shrimpfish's body, so it is sometimes hard to tell which is spine and which is shrimpfish. Juveniles are more often associated with sea urchins, while adults frequent beds of eel grass, where they are also effectively concealed.

No stomach?

Shrimpfish probably feed on small particles of plant and animal matter, and tiny planktonic crustaceans such as copepods. They suck the food into the mouth in the same unusual way as does the seahorse, using a mechanism equivalent to that of a pipette. Shrimpfish do not have a part of the gut easily identifiable as a stomach, and digestion is presumed to take place in the intestine. A surprising number of other fish have no stomach, the gullet passing straight into the intestine, and it is not always clear how and where the digestive juices are secreted or how they operate. There are also fish that have a stomach, but their digestive juices have not been detected; even the glands that secrete them seem to be absent. It is not only fish like the shrimpfish that lack stomachs. The skippers or saury pikes, *Scomberesox*, Atlantic fish related to flying fish and garfish, are also stomachless.

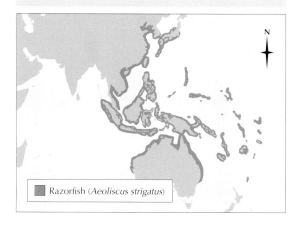

Razorfish (*Aeoliscus strigatus*)

SIDEWINDER

Also known as the horned rattlesnake, the sidewinder is named after its peculiar form of locomotion, which allows it to move over soft sand. Sidewinders are small rattlesnakes, the adults being only 1½–2½ feet (45–75 cm) long. The females are usually larger than the males, whereas in other rattlesnakes (discussed elsewhere in this encyclopedia) the reverse is generally the case. The body is stout, tapering to a narrow neck with a broad head that is shaped like an arrowhead. Above each eye there is a scale that projects as a small horn; a dark stripe runs backward from each eye. The body is pale gray or light brown, with a row of large dark brown spots running down the back and smaller ones on each side. The tail is marked with alternate light and dark bands, and the underparts of the body and tail are white.

The single species of sidewinder lives in the deserts of the southwestern United States, including Nevada, Utah, California, Arizona and in the northern part of Mexico.

Sand snake

Sidewinders are most common in areas of loose, windblown sand; although they can be found among rocks or on compacted sand, there is usually loose sand nearby. Although other rattlesnakes live in deserts and can be found on loose sand, the sidewinder is the most characteristic rattlesnake species in this type of habitat. It is likely that in such an environment the sidewinder has an advantage over other snakes. By adapting to life in moving sand the sidewinder does not compete with other snakes, which move over sand with an eel-like wriggling motion. The sidewinder's unusual looping movement enables it to get a good grip on loose sand and, consequently, to move faster.

Sidewinders are most active in the early part of the night, when air temperatures are not dangerously high and when their prey is also active. They spend the day in mouse holes or buried in sand, usually under the shelter of a creosote bush or yucca. They bury themselves by shoveling sand over themselves with looping movements of the body until they are coiled like springs and lie flush with the surface of the sand. Combined with their mottled brown coloration, this makes them very difficult to distinguish.

Desert prey

Shallow saucers are often visible in the sand near mouse and rat burrows in the desert. The depressions indicate places where sidewinders have rested; the snakes are probably attracted to these areas where they will find prey plentiful. Sidewinders feed mainly on small rodents, such as deer mice, kangaroo rats and spiny pocket mice, and lizards such as the tree-climbing utas and

The sidewinder's distinctive sideways movement enables the snake to obtain a good grip on the smooth sands of its native environment.

The sidewinder is able to move over soft sand without pushing against it, thereby keeping much of its body off the sand, which may be dangerously hot.

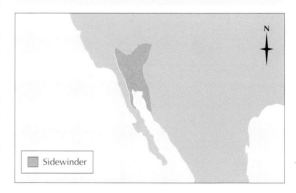

Sidewinder

other sand-dwelling iguanids. They also eat a few snakes, such as the glossy snake, *Arizona elegans*, other sidewinders and some small birds.

The breeding habits of sidewinders are the same as those of other rattlesnakes. Mating takes place when they emerge from hibernation in spring, and the young are born fully hatched.

Sidewinding

Many snakes perform sidewinding movements if they encounter a smooth surface, such as a sheet of glass, throwing their bodies into loops to get a grip on the smooth surface. As well as the sidewinder, the horned adder, *Bitis caudalis*, and some other snakes of southern African deserts also make a habit of sidewinding, leaving characteristic tracks in the sand. These are a series of parallel, wavering lines, each with a hook at one end made by the sidewinder's tail.

It is very difficult to see how the sidewinder makes its track without seeing the snake itself in action. In normal, or rectilinear, movement, a series of waves passes down a snake's body, pushing against the ground and driving the snake in the opposite direction to the waves. Sidewinding is very different movement, reminiscent of a coil spring being rolled or the movement of the tracks of a caterpillar tractor. The snake throws its body into curves, and, when moving, only two points of the body touch the ground. These points remain still while the raised parts move at an angle to the direction of the waves that pass along its body. As the snake

progresses, the part of the body immediately behind is raised so that the body is laid down and taken up in the manner of a caterpillar track. When the point of contact reaches the tip of the tail, a new point is started at the head end and the snake moves along a series of parallel tracks.

SIFAKA

THE SIFAKA IS ONE OF the more familiar lemurs on Madagascar, the island off the southeastern coast of Africa that harbors so many unusual animals. It belongs to the same family as the indri.

A sifaka has a short, black muzzle, which is naked, and forward-looking eyes that are golden in color. It has long, powerful hind legs for jumping. Its head and body length is about 18–22 inches (45–55 cm) and its tail measures about the same. Its long silky fur is sometimes white, but the various subspecies, about nine of them, have a variety of markings, which may be reddish, gray, yellow, chocolate brown or even black. In the north of the island there is a small blackish subspecies that is known as Perrier's sifaka.

Previously, zoologists thought that there was only one species of sifaka, which was highly variable in appearance. However, differences in social behavior and in their calls justify recognition of three species: Verreaux's sifaka, the diadem sifaka and the golden-crowned or Tattersall's sifaka. The last-named species, already on the brink of extinction, was only discovered and described by scientists in 1988.

Peaceful sun-loving lemurs

Sifakas live in small troops, averaging five per troop. These are not family groups but consist of several adults of both sexes, and generally there is only one infant at a time. Each troop defends its territory against neighboring troops. The animals live mainly high in the trees, at 40–70 feet (12–22 m) above the ground. They usually sit in a vertical position in a crotch of the tree, clutching the trunk. When on the move, they may hop along a horizontal branch or along the ground on their long hind legs, with their arms held high, out of the way. They rarely walk, either bipedally, or on all fours. Their legs are so long that they are much more at home hopping, or they may travel by swinging by their arms.

When it leaps, a sifaka turns its body, pushes off with its feet and launches itself into the air with its body held vertical, landing on the feet first, followed by the hands. Leaps of 33 feet (10 m) are known. In the dry spiny forests of the southwest, the main vegetation is the tall succulent *Euphorbia*, which sifakas negotiate without any apparent discomfort from the spines.

Sifakas spend a lot of time basking, often during the early morning. Each animal leans back, usually without hanging on, and spreads its arms and legs, its head lolling lazily to one side, letting the sun warm up the sparsely haired underside. When the front is done, the animal may turn around and do the back. Sifakas groom themselves, especially after sunning, by combing and scraping their fur with the forward-leaning lower incisors that are characteristic of lemurs. They finish by scratching with the long grooming claw on the second toe, another feature of all lemurs, and they may clean out their ears with this claw. They also groom each other. A pair, usually a male and a female but sometimes two males, sit facing each other and groom with the teeth and tongue, not with the claw. After grooming, they may spar with their hands and feet, but this is probably only play, since relations are peaceful within the troop, and, except in the breeding season, there are no fights or even a show of fighting.

Verreaux's sifaka is one of the more common lemurs in Madagascar. It is still considered endangered, however, because its habitat is dwindling so fast.

On rare visits to the ground, Verreaux's sifaka uses a unique bipedal hopping movement. Its long hind legs make walking awkward.

Sifakas have superb stereoscopic vision and at least some degree of color vision. Their sense of smell is also acute. A sifaka troop will often try to invade another's territory. Keeping close together and moving cautiously, the invaders hop into the forbidden zone and, if not observed by the residents, they fan out to feed. When the residents see them, they hop toward them and the invaders hop backward. Each individual remains facing the direction he started with, hopping forward, backward or sideways; two opponents may even end up back to back. Although the invaders always retreat, constant conflicts, day after day, may ultimately result in a change in the territorial boundary.

Mainly fruit eaters

All members of the Indriidae family are specialized leaf eaters. They have strong chewing muscles, a complex stomach and a very long intestine, 10–15 times the body length. It seems, however, that sifakas eat leaves somewhat infrequently. Their food is only 25 percent leaves and 65 percent fruit, the other 10 percent being made up of flowers and seeds. No meat or insects of any kind are eaten. A sifaka pulls a twig or branch toward itself with its hands, but it always plucks the food with its teeth. As a result, much of the food is clumsily dropped. Even on the

---

SIFAKAS

CLASS	**Mammalia**
ORDER	**Primates**
FAMILY	**Indriidae**

GENUS AND SPECIES **Diadem sifaka, *Propithecus diadema*; Verreaux's sifaka, *P. verreauxi*; Tattersall's sifaka, *P. tattersalli***

ALTERNATIVE NAME
Golden-crowned sifaka (*P. tattersalli*)

WEIGHT
11–18 lb. (5–8 kg)

LENGTH
Head and body: 18–22 in. (45–55 cm); tail: 18–22 in. (45–55 cm)

DISTINCTIVE FEATURES
Forelimbs shorter than hind limbs; naked, black face, palms and soles; thick, silky fur on back; fur on back of ears; rounded head and ears; much color variation: white, yellow, grey or black upperparts, with contrasting underside

DIET
Fruits, leaves and shoots; some bark and dead wood

BREEDING
Age at first breeding: 2–2½ years; breeding season: January–March; number of young: 1; gestation period: 120–160 days; breeding interval: 1–2 years

LIFE SPAN
At least 7 years

HABITAT
Dry, deciduous and evergreen forests

DISTRIBUTION
Close to east and west coast of Madagascar

STATUS
Vulnerable: *P. verreauxi*; critically endangered: *P. d. perrieri, P. d. diadema* and *P. tattersalli*

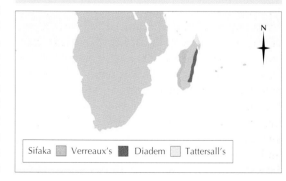

Sifaka	Verreaux's		Diadem		Tattersall's

ground sifakas do not use their hands. Sifakas seem not to drink and apparently get all the required moisture from their food.

Baby worship

The mating season is from January to March. In spite of their otherwise peaceful disposition, the males may fight at this time, during which they may sustain slight injuries such as torn ears. Only the winners of these fights get the opportunity to breed. The single baby is born after a gestation period of 130 days. It is carried on the mother's belly from July, which is the peak month for births, until October, when it transfers itself to her back and is carried there for the next 2 months. Sifakas reach full size at 21 months but the males are not sexually mature until 2½ years of age. A female with a baby becomes the focus of attention for the rest of the troop, who crowd around and try to groom the infant. As a rule, sifakas do not do well in captivity but a few have lived for over 20 years in zoos.

Singing their troubles away

The main defensive tool of a sifaka seems to be its voice. Its main predator, as for all lemurs of Madagascar, is the harrier hawk, genus *Polyboroides*. On seeing one of these, a sifaka gives the alarm call, a roaring bark made with the head held back and the mouth shaped like an O. It is a very resonant call and the whole troop joins in in unison. Other troops in the neighborhood, hearing the cry, look nervously at the sky, and if they see the hawk, they all take up the roar. On the ground the chief predator is the civetlike mammalian carnivore, the fossa, *Cryptoprocta ferox*. When mobbing the fossa, sifakas utter the cry that gives them their name: *si-fak*, a sound that has been described as a bubbling groan made with the mouth shut and ending with a sharp open-mouthed cluck. A troop will go on calling for up to 45 minutes, and if the fossa does not then go away the troop retreats. Sifakas use the same call toward a human, and the *si-fak* chorus is said to be quite unnerving.

Disappearing lemurs

Animal species have been becoming extinct due to human activities for years. On Madagascar, where much of the animal life is unique to the island, the folklore includes references to animals that the people's ancestors knew but that do not exist today. One of these, called tratratratra, is based on a creature not long extinct. E. de Flacourt in 1661 gave an accurate description of the tratratratra, saying it was the size of a two-year-old calf with a round head, a humanlike face and monkeylike feet. There is firmer evidence from the fossils that have been discovered in

Madagascar. Radiocarbon dating shows that they are not more than 1,000–2,000 years old. They were probably exterminated by humans. Many extinct lemurs were larger than those seen today, some being the size of small orangutans, and they probably moved through the trees in the same leisurely fashion as orangutans. Somewhat different were *Archaeolemur* and *Hadropithecus*, which, with long slender legs, were probably fast runners and lived on the ground. The largest of all the extinct lemurs was *Megaladapis*, up to 4 feet (1.3 m) long in head and body, with a long, massive face, small braincase and robust skeleton. Lemurs would have been easy enough for even primitive man to kill. One skull of *Archaeolemur* has a dent made by an ax during the animal's lifetime, and the bones of extinct lemurs show cut marks and signs of burning.

Tragically, at the present rate of deforestation and habitat destruction on Madagascar, the sifakas and other lemurs seen today will soon be joining the tratratratra, and will exist only in legend, folklore and history. All subspecies of sifakas are already listed by the I.U.C.N. (World Conservation Union) as endangered.

Verreaux's sifaka prefers to travel from tree to tree at canopy height. Due to its tree-top lifestyle, drinking water is hard to come by, but it survives on the moisture in its food, and from dew licked from its body each morning.

SIKA

Sika favor acid soil habitats, such as forests of pine and Sitka spruce. The light-colored band on the faces of the hinds (females) in the picture above is characteristic of both sexes.

THE SIKA IS CLOSELY RELATED to the wapiti or red deer, *Cervus elaphus* (discussed elsewhere). It is a little smaller than the fallow deer, *Dama dama*, to which it is also related, and stands up to 32¾ inches (82 cm) at the shoulder. The pure white tail enables it to be easily distinguished from a fallow deer, in which a black tail contrasts with a white rump. In both deer the rump is bordered with black. The faint spotted coat is buff brown, becoming darker in winter; the head is a paler brown. Along the neck and back there is a black stripe. The male's antlers develop two short spikes in the second year, but never have more than five tines (points) each. The antlers start to grow in May, the velvet is shed in August and the antlers are cast the next April.

The original home of the sika is in the far east of Asia, in China, Taiwan, Manchuria, Korea and Japan. They are now very rare in most of this area but flourish in many parts of Europe, where they have been introduced. Originally kept in parks, many sika escaped, creating feral herds in Austria, Denmark, France, Germany and Britain. Many thousands of sika also live in preserves in the former Soviet Union.

Small herds

Sika live in deciduous or mixed forests as well as grassland and marsh. They are very hardy, and are able to withstand severe frosts, but their range is limited by snowfall. In the summer they range widely through forests but in winter they gather in places where the snow cover is scanty. Sika are less gregarious than red deer and, except during the rut and in winter, when small groups form, both stags and hinds live solitary lives.

Sika feed at night, coming out of the thick cover where they have spent the day to browse and graze on grassland, moor or open woodland. In the summer they eat grass and coarse herbage and also pluck buds and leaves from trees and bushes. In the winter they strip bark and eat twigs when better food is not available. They occasionally feed on green algae on the seashore. Sika move continuously while they feed, slowly wandering from one plant to another, tearing off the tender parts and leaving the rest.

Whistle-grunt calls

The rut (mating season) starts in September, when the antlers have shed their velvet, and lasts for 6–8 weeks. During this period the males issue calls, a characteristic rising and falling whistle that ends in a grunt, repeated three or four times. Males fight each other for the opportunity to gather a harem of females.

The calves are born the following May or June and they resemble the fawns of fallow deer. About 24 hours before the single calf, rarely twins, is born, a female sika leaves the other

SIKA

CLASS	**Mammalia**
ORDER	**Artiodactyla**
FAMILY	**Cervidae**
GENUS AND SPECIES	***Cervus nippon***

WEIGHT
55–242 lb. (25–110 kg)

LENGTH
**Head and body: 38–56 in. (95–140 cm);
shoulder height: 25½–32¾ in. (64–82 cm);
tail: 3–5 in. (7.5–13 cm)**

DISTINCTIVE FEATURES
**Chestnut coat, often covered in white spots
across back; lighter underside and legs;
white tail; dark mane on neck in winter
(both sexes); erect antlers with 2 to 5 tines
(points) on each (male only)**

DIET
Grasses; occasionally shoots and leaves

BREEDING
**Age at first breeding: about 18 months;
breeding season: September–October;
number of young: usually 1; gestation
period: about 30 weeks; breeding interval:
about 1 year**

LIFE SPAN
**Up to 15 years; perhaps up to 25 years
in captivity**

HABITAT
Forests, grasslands and marshes

DISTRIBUTION
**Eastern Asia; Southeast Asia; Japan;
introduced to many parts of Europe**

STATUS
**Several subspecies classed as endangered,
some may already be extinct; locally
common in introduced range**

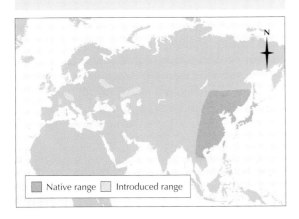

Native range Introduced range

females and retires to a thicket. The calf is able to walk in a few hours and can run well in a couple of days, but spends most of its time lying in cover. Weaning takes place at 8–10 months.

Competition with other deer

In Britain the spread of feral sika from deer parks appears to have been detrimental to the population of roe deer, *Capreolus capreolus*. Where the two exist together, the roe deer become rarer and in some places have even disappeared. For some reason, the two species compete against each other, as do roe deer and fallow deer. An exception is where the sika prefer one habitat, such as damp woodland, and the roe prefer another, for example dry heathland.

In China, sika compete with red deer in a different way. Male red deer sometimes drive sika away and mate with their females to produce hybrid deer, similar to sika but larger.

A sika stag's tines become whiter during the winter. The antlers are widely used in traditional Chinese medicine.

SILK MOTH

Silk moths in domestication

In the silk industry the larvae are called silkworms and the eggs are called seed. The eggs are stored during winter at a low temperature. As soon as mulberry leaves become available, the eggs are removed to surroundings at a temperature of 65° F (18° C), which is slowly increased to about 77° F (25° C). After they hatch, the larvae are kept at this temperature and feed for 42 days. They change their skins four times, each skin change involving 24 hours of inactivity when the larvae do not feed. When fully grown, the larvae stop feeding altogether and begin to make side- to-side weaving movements with their heads. They are then given bushes to climb up and they spin their cocoons among the twigs. The cocoons are then exposed to steam or hot air to kill the pupae within before the silk is reeled.

An ounce of eggs or seed produces about 30,000 larvae, which generally produce 12 pounds (5.5 kg) of silk, consuming 1 ton (0.9 tonne) of mulberry leaves in doing so. A silkworm increases its weight about ten-thousandfold from when it hatches to the time it is fully grown.

Over time, silk manufacturers have selectively bred silk moths to aim for the production of larger cocoons, better silk and more docile and manageable insects. The silkworms now have no instinct to wander or hide themselves but can be kept in open trays. (By contrast, the larvae of all wild moths must be carefully caged in captivity.) Silk moths have well-developed wings, but they cannot fly. Domestic silk moths of this kind have existed in China for hundreds of years. In the early period of domestication there must have been frequent accidental pairings of wild moths with escaped or discarded domestic specimens. As a result the wild stock probably became degenerate through genetic contamination of characteristics useful in a domesticated insect but dangerous or fatal in a wild one. However the ancestral silk moth may survive in some parts of central Asia, remote from the Chinese silk-producing areas.

Other silk moths

None of the other species of *Bombyx* has been used with any success for silk culture, but several moths quite unrelated to the mulberry silk moth

The cultivated silk moth, inbred due to many hundreds of years of commercial production, has lost many of the behavioral habits it may have had in its natural state.

THE SILK MOTH USED in the commercial production of silk is the mulberry silk moth, *Bombyx mori*. It is a medium-sized, light-colored moth of eastern Asian origin. Related species, belonging to the family Bombycidae, are found in eastern Asia and in Africa, although there are none in Europe. The mulberry silk moth is the only completely domesticated insect; the species is now entirely unknown in its wild state. None of the many strains or races of domestic silk moths could possibly fend for themselves or exist in the wild, although other species of silk moths are still present in the wild.

The natural food of the silk moth larva is mulberry leaves. When it is ready to pupate the larva spins a thick, strong silken cocoon formed from a single continuous filament of pale yellow silk about 1,000 feet (300 m) long, which can be spun or reeled off as a single strand. The strains of silk moth that appear to be nearest to the ancestral form have one generation a year, passing the winter as eggs.

SILK MOTH

PHYLUM	**Arthropoda**
CLASS	**Insecta**
ORDER	**Lepidoptera**
FAMILY	**Bombycidae**
GENUS AND SPECIES	***Bombyx mori***

ALTERNATIVE NAME
Silkworm (larva only)

LENGTH
Adult wingspan: 2 in. (5 cm); larva up to 3 in. (8 cm)

DISTINCTIVE FEATURES
Moths: pale yellow-cream color; rounded furry bodies; pale wings with obvious veins; slightly hooked forewings. Larvae: pale with pinkish spots on back.

DIET
Larvae: mulberry leaves

BREEDING
Female lays about 400 eggs on mulberry plant; hatch into larvae; spin cocoons and pupate

LIFE SPAN
Variable, perhaps 1 year or more

HABITAT
Not known in wild state

DISTRIBUTION
Originally eastern Asia, now commercially managed

STATUS
Common in commercial environment

and belonging to the emperor and atlas moth family Saturniidae have formed the basis of minor silk industries in India, China and Japan. The most important of them are the producers of tussore and muga silk, large moths of the genus *Antheraea*. These include *A. pernyi*, *A. mylitta* and *A. paphia* and the Japanese species *A. yamamai*. Another silk producer from India is the Eri silk moth *Attacus ricini*, a relative of the giant atlas moth. The Indian silk moth, *Theophila religiosae*, is also commercially bred for its silk.

In recent years these moths have become an object of interest to dealers in Europe and North America, who import them and sell the eggs or pupae to naturalists to rear for the pleasure of seeing the moths and their larvae. Locally available food plants can generally be found for them; most species of *Antheraea* will feed on oak.

Many other exotic Saturniids are imported and all are known collectively as giant silk moths, although most of them have never been connected in any way with the production of usable silk, and they bear no affinity or resemblance at all to the true silk moth.

There are 300 species of moths in the family Bombycidae. The pale larvae of each species spin silken cocoons before they pupate.

Keeping silk moths

Silk moths are easier to manage than other moths and butterflies because of their inbred docile nature. The eggs are best obtained in the spring, when the mulberry tree comes into leaf. If they are purchased before this time, they should be kept in a refrigerator until the time is convenient for them to hatch. Then the tiny caterpillars should be put in a cardboard box lid or tray made of newspaper with mulberry leaves cut into strips; this provides more leaf edges from which the larvae can feed. They do well at room temperature but grow rather more slowly than commercially reared stocks. Slightly wilted lettuce leaves can be used as silkworm food, but they do not thrive as well as on mulberry leaves and, if started on mulberry leaves, do not take to lettuce.

When a larva stops feeding and stays in one place on a leaf or the bottom of the tray, it is preparing to shed its skin. If it is disturbed or moved during this period, which lasts about 24 hours, it probably will die. It is well worth keeping the caterpillar under constant observation to watch the actual process of skin shedding, which takes only a few minutes.

When they reach full size, about 2 inches (5 cm), the larvae move into corners and weave their heads about with a searching movement.

Originating from East Asia, the silk moth is now unknown in a natural state. It is generally maintained in commercial silk farms.

Each one should be put separately in a rolled and pinned paper cone and it will then proceed to spin its cocoon. After about 10 days the cone should be gently shaken; if it rattles, there is a pupa inside the cocoon. This is the time to reel the silk off. A cocoon from which the moth has hatched is useless because the moth cuts its way out of the cocoon at one end and so severs the continuous thread of silk in many places.

To reel the silk, the cocoon should be taken and all the loose outer fuzz pulled off until the cocoon is smooth and firm. It should then be soaked in warm water for half an hour. Then the silk should be gently pulled away from the sides a few strands at a time until a single strand comes loose. This is the continuous thread, which can be wound onto a reel or a piece of cardboard. It is best to reel the silk from several cocoons at once as this gives a bigger yield. The cocoons should be left to tumble about in the water as the thread is wound. Not all the silk will come off because the first spun, innermost part of the thread is more firmly glued and felted together than the rest of the cocoon. The end result should be a loop, or skein, of lustrous pale yellow silk. Treated in this way, the pupae are not injured and can be kept until the moths hatch from them.

The history of silk

The silk industry originated in China, and the traditional story is that it was first given official recognition by the wife of the Emperor Huang Ti in about 2140 B.C.E. The secret appears to have been guarded for over 2,000 years, but in the third century C.E. it reached Japan by way of Korea. A little later cultivation of silkworms was established in India, and because it started in the northeast, in the Brahmaputra valley, it seems likely that knowledge of the technique of silkworm cultivation came overland, direct from the Chinese Empire.

Around the beginning of the Common Era, silk from Asia was one of the most costly items in the trade between the Roman Empire and the Orient. The Byzantine emperor Justinian (c. 482–565) reserved the monopoly of the silk trade for himself and sent two Persian monks who had lived in China to bring the means of producing silk back to to Constantinople. They succeeded in doing so, smuggling the eggs in a bamboo tube in C.E. 550. From that time on, the silk industry gradually became established in Europe. It was widely spread by the eighth century and the conquering Arabs or Moors carried silk manufacture with them in the wake of their conquests both east into Asia Minor and to western Europe. France and Italy have always been the leading silk producers in Europe.

In recent years natural silk production has suffered severely from competition with artificial fibers. However, natural silk production is kept alive in the world, especially in Japan, by a combination of both hard work and efficient factory production methods.

SILVERSIDE

SILVERSIDES ARE MEMBERS of a family of around 150 species of small, slender-bodied fish called the Atherinidae. Silverside can refer to the species *Atherinopsis californiensis*, found off California and also known as the jacksmelt, but it can also mean atherinids in general. Silversides in the general sense are shoaling fish of inshore seas, and are frequently found together in large numbers. Many species are considered good eating, although small, and some species are fished commercially. Perhaps because of this, and their worldwide distribution, they go under a variety of names in different parts of the globe. For instance, the Samoan silverside, *Hypoatherina temminckii*, is called whitebait in South Africa, but whitebait refers to several different, unrelated species in such places as the United States and Britain. A general alternative name for silversides is sand smelt, not to be confused with the unrelated true smelt, a name used primarily in European waters. The silverside family also includes the grunion, described elsewhere.

The two European silversides are marine and not more than 6 inches (15 cm) long. They both have a slender body, small head with large eyes, two dorsal fins and an anal fin. The pelvic fins are set well forward. The back is green and the scales have black points around their edges. The underside is silvery white, with a brilliant silver band separating it from the green of the back. In North America the inland or tidewater silverside is a marine species, but it tolerates brackish water and has the ability to travel far inland to make its home in fresh water, in sheltered parts of large rivers and around the edges of large lakes, where it breeds. For this reason, it has been introduced in several parts of North America, notably large reservoirs, where it serves as food for sport fish.

Marine and freshwater types

There are entirely freshwater species in Central America, up to 20 inches (51 cm) long, that are caught for market, and two species in Lake Sentani, in New Guinea, which are also caught for food. The freshwater silversides of Australia are known as rainbowfish and are placed in a small family of their own, the Melanotaeniidae. Like the Cuban glassfish, *Alepidomus evermanni*, another species of silverside, they are exported and kept by aquarists.

Marine silversides live in compact shoals in coastal and inshore waters, swimming with the tip of the snout just below the surface. They are attracted to the less salty waters of estuaries,

docks and the saltings of saltwater marshes. When alarmed, the shoals make rapid changes of direction. Occasionally, a shoal may be left behind in one of the larger rock pools by the ebbing tide. In the fall the shoals move into deeper water, returning inshore in spring.

They feed on small plankton, especially small crustaceans and other food such as small marine worms. They also eat fish larvae and small fish. Silversides have very small teeth in

Here, the slender European marine silversides form a mixed shoal with a group of fuller-bodied chromis, a kind of damselfish.

The banded rainbow-fish, Melanotaenia trifasciata, of Australia, is a typical example of the rainbowfish family, a family recently separated from the silverfish family. It lives in small streams and waterholes, and often lingers around submerged logs and branches.

their jaws, on the palate and in the throat, reflecting the combination diet of plankton, for which they do not need teeth, and fish, for which they do. They eat plankton in summer when this is plentiful and small fish in winter when plankton is scarce.

Clinging eggs

Spawning takes place in summer. The eggs are around 2 mm in diameter and have a number of slender filaments that catch onto seaweeds and animals that are encrusted on seabed pebbles. The fry are ¼ inch (6 mm) long when hatched and at ½ inch (13 mm) long they form compact shoals that can often be seen swimming just below the surface at the water's edge. They grow rapidly, and at 1 year of age the silverside is about 2½–3 inches (6–8 cm) long.

Caught by the shoal

The flesh of silversides is very palatable. In New Guinea bundles of brushwood are put into shallow water. The silversides swim into these for shelter and are trapped when a net is placed around the brushwood. Apart from man, predatory fish and birds also take advantage of the silversides' shoaling behavior. Like other animals that aggregate, each individual silverside in a shoal actually lowers its chances of getting caught by a natural predator. A large shoal is conspicuous and easy for predators to locate, but this disadvantage is outweighed, from the point of view of each individual fish, by the low chance of any one fish being caught. Perhaps paradoxically, it is in the interests of every individual fish to shoal. Shoaling behavior is not a good defense against humans, who have developed techniques to catch entire shoals at once.

SILVERSIDES

CLASS **Osteichthyes**

ORDER **Atheriniformes**

FAMILY **Silversides, Atherinidae; rainbow fish, Melanotaeniidae**

GENUS AND SPECIES **Atherinidae includes inland silverside, *Menidia beryllina* (detailed below) and Peruvian silverside, *Odontesthes regia*. Melanotaeniidae includes MacCulloch's rainbowfish, *Melanotaenia maccullochi*.**

ALTERNATIVE NAME
Menidia beryllina: **tidewater silverside**

LENGTH
Up to 6 in. (15 cm)

DISTINCTIVE FEATURES
Slender body; 2 dorsal fins; scales on back outlined in black; silver bands on sides

DIET
Animal plankton; also insect larvae in fresh water

BREEDING
Breeding season: spawns in March–September in Florida but in April–June in New York; number of eggs: 15,000; hatching period: 4–30 days

LIFE SPAN
Up to 1 year

HABITAT
Inshore waters of sea, estuaries, large rivers and littoral zone of lakes. Near surface of clear, quiet water, over sand or gravel.

DISTRIBUTION
Western Atlantic from Massachussetts to Florida; Gulf of Mexico to Yucatan. Inland, through Mississippi River system to southern Illinois and eastern Oklahoma. Introduced to landlocked locations in U.S.

STATUS
Locally common

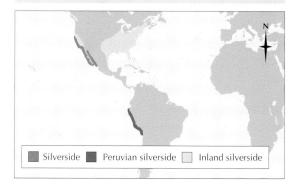

Silverside ■ Peruvian silverside ☐ Inland silverside

SIREN

IRENS ARE NORTH American amphibians named after mythical beings. They are related to salamanders but look like eels, and are sometimes called mud eels. Sirens never properly metamorphose into an adult form, retaining larval features throughout life.

The greater siren is up to 3½ feet (1 m) long and is slimy to the touch. It is dark gray-green with its underside flecked with green or yellow. The body is long and slender, tapering more or less evenly to the tip of the tail. Behind the small head, on either side, are bunches of feathery external gills. The siren has no hind legs. It has even lost all trace of the bones of the hip girdle. Its front legs have four small toes, but are small and weak.

The greater siren ranges from Washington, D.C., to Florida. The lesser siren ranges over the southeastern United States, the lower Mississippi Valley, parts of Texas and into Mexico. It measures 7–20 inches (18–51 cm) and is dark brown or bluish black. There are also two species of dwarf sirens: the northern dwarf siren, found in from Georgia to Florida, and the southern dwarf siren, found only in Florida. Each is up to 10 inches (25 cm) long and drab colored, with three toes on each forefoot, and the forelegs are even smaller than in the larger species. Although these species are very similar and found together in Florida, they differ in their chromosomes and in their choice of microhabitat and diet.

Surviving the drought

The greater and lesser sirens live in shallow streams and ditches, hiding by day among water plants and coming out at night to feed. They swim by wriggling their bodies in an eel-like way. If they find themselves out of water on mud, the legs may be used to drag themselves along. The dwarf sirens are restricted to marshes, swamps and bogs, burrowing in mud or hiding among submerged vegetation. Sirens are almost toothless but have teeth in the front part of the roof of the mouth. The jaws have horny plates instead of teeth. They eat both plants and aquatic invertebrates. The two larger sirens take mainly invertebrates such as worms, water snails, freshwater shrimps and crayfish, but also eat green filamentous algae. A large part of the diet of the southern dwarf siren is the leaves of the water hyacinth, genus *Eichhornia*.

Sirens keep their external gills throughout life. In periods of drought they burrow up to 1 foot (30 cm) into the mud. Their skin loses its slimy texture, their gills become smaller, sometimes almost disappearing, and they enter a state of dormancy or suspended animation. These dry conditions may last for up to two months with little harm to the sirens. Then, when the rains come and the streams and swamps fill up with water, the sirens' gills return to their normal size. The body becomes slimy and the sirens resume their usual activities.

The lack of hind limbs is just one of the range of unusual characteristics that distinguish sirens, such as this lesser siren, from other amphibians, and leads some zoologists to place them in their own suborder, separate from salamanders.

Lesser sirens, like other sirens, possess branching, feathery external gills through-out life, instead of metamorphosing and losing them as do most other salamanders. They use a horny beak for feeding and are almost toothless.

A peculiarity of sirens is that they have very few ribs. Because the body is so long in relation to its girth, the pumping capacity of the heart is increased. This is achieved by the auricles of the heart, which have fingerlike extensions from their sides that increase their capacity.

Sirens have no poison glands in the skin to protect them, as so many amphibians have, and their chief predator is the red and black rainbow snake, *Abastor erythrogrammus.*

Amphibious Peter Pans

The female sirens lay their eggs singly or in batches on the leaves of water plants. The development of the larva follows the usual lines of salamander and newt larvae except that there is no metamorphosis to the adult. The adults remain physically like the larval form. While keeping larval features such as their external gills, they become sexually mature, a process known as neoteny (see axolotl). So the sirens never grow up. As they live wholly aquatic lives, they seem to have dispensed with the terrestrial adult stage of the life cycle during their evolutionary history, just as they have dispensed with their unnecessary hind limbs.

Confusion confounded

Why these animals should be called sirens is hard to say. The sirens of Greek mythology were water nymphs who lived on the rocks by the shore and sang sweetly to lure mariners to their doom. The sirens of the streams and marshes of North America make no more than a faint whistle or a tiny yelp. The choice of name is unfortunate because the sea cows, large marine mammals of order Sirenia (see "Manatee" and "Dugong"), are also sometimes called sirens.

SIRENS

CLASS	**Amphibia**
ORDER	**Caudata**
FAMILY	**Sirenidae**

GENUS AND SPECIES   **Northern dwarf siren, *Pseudobranchus striatus*; southern dwarf siren, *P. axanthus*; lesser siren, *Siren intermedia*; and greater siren, *S. lacertina***

ALTERNATIVE NAME
Mud eel

LENGTH
***S. lacertina*: 20–39 in. (50–98 cm).**
***P. striatus*: up to 10 in. (25 cm).**

DISTINCTIVE FEATURES
Long, cylindrical, eel-like body; small forelimbs; hind limbs and pelvic girdle absent; prominent external gills; small eyes without eyelids

DIET
Aquatic plants and invertebrates

BREEDING
Age at first breeding: 2 years; breeding season: February–March; fertilization external; number of eggs: 200; hatching period: 4–8 weeks; breeding interval: 1 year

LIFE SPAN
Up to 25 years in captivity

HABITAT
Fully aquatic; shallow streams, ditches and other still water bodies with abundant vegetation and muddy bottoms

DISTRIBUTION
Southeastern U.S., especially coastal plains of South Carolina, Georgia, Florida, Louisiana and Mississippi; also Mexico

STATUS
***Vulnerable (S. intermedia)* to not threatened (other species)**

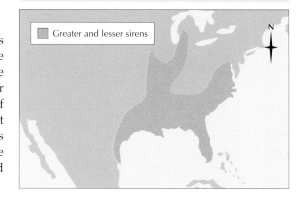

Greater and lesser sirens

SKATE

SKATES ARE FLATTENED FISH related to sharks. There are around 14 genera and more than 200 species in the skate family, the Rajidae. Broadly speaking, a skate is a type of ray, and some species assigned to the skate family are informally called rays. For example the thornback skate, *Raja clavata.*, is sometimes called the thornback ray. However, there are a number of features of skates that set them apart from rays, and experts classify these fish as either skates or rays. The original name, skate, dating from the 12th century, referred to one species, *Dipturus batis*, formerly the common skate. Ironically, the common skate has been depleted by overfishing, becoming extinct in such areas as the Irish Sea, after a taste was acquired for skate in Britain at the start of the 20th century. The alternative name of blue skate now seems more appropriate.

The body of skates is flattened, and the pectoral fins are so large that they dominate the body, giving it an overall rhomboid, diamond or disk shape when viewed from above. The tail is short and stout in comparison to rays. It has a row of spines running along the mid-dorsal line and two small dorsal fins near its tip, as well as a tiny, reduced caudal (tail) fin. The front of the head ends in a snout, which is short in some species and long in others. The eyes are on top of the head, with an inlet for water, called a spiracle, behind each. The mouth and nostrils are on the underside, as are the five pairs of gill slits. The pelvic fins are small and in males the inner edges carry a pair of so-called claspers, which are organs for placing the sperm inside the female. The common skate grows to 6 feet (1.8 m) long or more, whereas the thornback skate grows to 3 feet (0.9 m). One of the largest is the big skate of the American Pacific, *R. binoculata*, which grows to 8 feet (2.4 m).

Skates and rays are found in most temperate and tropical waters except much of the South Pacific and an area off northeastern South America. They live on continental shelves, often to depths of 600 feet (180 m). The deep-sea skate, *Bathyraja abyssicola*, lives down to 9,530 feet (2,905 m) off the Pacific coasts of North America and Japan.

Food enveloped

Skates tend to live on sandy, gravelly or muddy bottoms, spending most of their time lying on the seabed. Occasionally, they might swim upwards with a wavelike flapping of the pectoral fins, often called *wings*, the free edges of which continuously undulate. The tiny caudal fin towards the tip of the tail plays no significant role. Skates migrate back and forth into deeper water with the changing seasons. The common skate moves into deeper water for the winter, whereas the thornback moves into shallow water. All skates and rays are bottom-feeders. As

The diamond shape of this thornback skate is formed during development by the outsized pectoral fins. The spiracles are clearly visible in this picture, as are the two small lobes of the pelvic fins.

When found washed up on a beach, the skate's egg capsule, or "mermaid's purse," lacks both its amber color and the feltlike fibers that cover it when the young skate is developing inside.

young fish they take small crustaceans such as half-grown shrimps and other small invertebrates. Later they eat bottom-living animals such as crabs and lobsters as well as some fish. The larger individuals of the common skate eat gurnards, flatfish and anglerfish, as well as midwater fish such as herring, scad and pilchard. Small squids are sometimes eaten. Prey is taken with a pounce, the skate smothering its victim with its wide pectoral fins and then seizing it with its mouth. Although skates seem to have good sight, they locate their prey by scent, and probably also by passive location with their electroreceptors, so they can hunt equally well by day or night.

Specialized breathing

When a fish spends so much time on the bottom with its mouth and gills on the underside, there arises a special problem: its breathing. A fish normally breathes by actively passing water over the gills. The water is taken in through the mouth, passes through the gill cavity and then back out to the exterior. Sharks and rays make water flow in this way by swimming forwards, but when a skate is lying on the bottom, its mouth is closed and apparently there is no water movement. However, water is taken in through the spiracles on the top of the head, then passes into the gill cavities and out through the gill slits on the underside. By using this arrangement, the skate also avoids swallowing sand or mud.

Babies in purses

Some time after mating, the female lays her eggs. The eggs are each enclosed in an unusual structure, a rectangular capsule with a hornlike process at each corner, which in some species is

COMMON SKATE

CLASS	**Chondrichthyes**
ORDER	**Rajiformes**
FAMILY	**Rajidae**
GENUS AND SPECIES	*Dipturus batis*

ALTERNATIVE NAMES
Blue skate; gray skate

WEIGHT
Up to 213 lb. (97 kg)

LENGTH
Up to 8⅓ ft. (2.5 m)

DISTINCTIVE FEATURES
Disc-shaped body; wings sharply angled with front edge slightly concave; skin rough over whole surface (male), on front region (female) or smooth all over (juvenile); tail with central row of spines

DIET
Large range of bottom-dwelling animals; fish preferred by larger individuals

BREEDING
Age at first breeding: mature at length of 20 in. (50 cm); breeding season: March–September; number of eggs: 6 to 40

LIFE SPAN
Not known

HABITAT
Seabed in coastal waters on sand and mud at depths of 100–2,000 ft. (30–600 m); sometimes in very shallow water

DISTRIBUTION
Eastern Atlantic Ocean: from Norway, Iceland and the Faeroe Islands to Senegal, western Mediterranean and western Baltic

STATUS
Endangered: extirpated from much of its former range by trawling

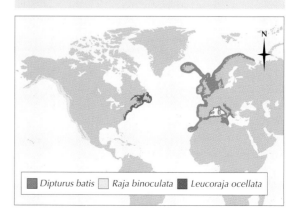

Dipturus batis ☐ *Raja binoculata* ■ *Leucoraja ocellata*

drawn out into a tendril. The tendrils help to anchor the egg cases to objects on the sea floor. In places the surface of the capsule is covered with a felt of loose fibers. In the common skate the fibers near the corners and at the bases of the processes are long filaments. The capsules differ in size according to species, those of the thornback skate being 3–4 inches (7.5–10 cm) long, whereas those of the common skate are more than double this size. The baby skate, when it hatches, is almost as broad across the pectoral fins as its capsule is long.

The capsules are amber-colored at first but turn black after the young have hatched. As the capsules are rolled about on the seabed, the fibers are rubbed off. The smooth, black egg cases of the thornback skate are often washed ashore and are called mermaids' purses.

Winged food

Skate is a commercially valuable fish in European fisheries but not so in the United States. In Britain, the common skate was once despised. The development of the fishery dates roughly from the beginning of the 20th century. Skate are taken in trawls and on longlines, as well as on rod and line, with which fish of up to 213 pounds (97 kg) have been landed. The skate fishery has now moved to deeper waters, having depleted the common skate through much of its former range on the continental shelf. The wings, or pectoral fins, are the parts used for food.

Electric mimic?

Skates, in common with sharks and rays, have arrays of electroreceptors around their snouts, which they use to detect prey by the electricity in the prey's muscles. More unusually, skates also exploit biological electricity by generating it themselves using electrogenic organs developed from modified tail muscles. There are several differences from the electrogenic organs of electric rays. While electric rays generate a high voltage that they can use to stun prey, those of the skates generate less than 4 volts, a much lower charge.

The exact function of the skates' electrogenic organs is not yet certain, it but has been the subject of fascinating speculation. The skates may use their electric discharges to confound their predators, such as the monkfish or angelshark, (genus *Squatina*), which, like the skate, hunts by means of its electroreceptor organs. By generating electric discharges, the skate may mimic the muscle activity of a much larger and more fearsome animal. Alternatively, the discharges might simply confuse the angelshark by disrupting its electroreceptive sense information.

The false eyes on the wings of this big skate probably deceive predators, giving misleading visual information about its size. The electric discharges of some skates may complete the illusion of a much larger animal.

SKIMMER

The black skimmer's unique feeding technique is tactile; it does not seem to see its prey before its bill makes contact. This means it can continue to hunt even on the darkest nights.

THE SKIMMERS ARE RELATED to the gulls and terns, but not as closely as was once thought. They are remarkable for their bills, of which the lower half is considerably longer than the upper half. They are also notable for their catlike eyes with vertical pupils, which are unique among birds. Apart from the bill, they resemble terns in appearance. They are 15–20 inches (38–51 cm) long but more heavily built then terns. The wings are very long and narrow, the tail is quite short but forked and the legs are also short. The upperparts are black or brown and the underparts are white. The sexes are alike in plumage, but the females are significantly smaller, by some 10–25 percent.

The black skimmer is the largest of the three species of skimmers, and ranges along the eastern coast of North America from New Brunswick southward and down both coasts of South America to Argentina and Colombia. The African skimmer is found from Senegal and Egypt to South Africa, and the Indian skimmer lives in India and Indochina.

Scooping fish

Skimmers feed by using their unique bill as a scoop to catch fish while in flight. Skimmers occasionally wade in shallow water, but the usual way of feeding is to skim rapidly over the surface of the water with the long lower half of

SKIMMERS

CLASS	**Aves**
ORDER	**Charadriiformes**
FAMILY	**Rynchopidae**

GENUS AND SPECIES **Black skimmer,** *Rynchops nigra* **(detailed below); Indian skimmer,** *R. albicollis*; **African skimmer,** *R. flavirostris*

WEIGHT
Male: 12⅓ oz. (350 g); female: 9 oz. (250 g)

LENGTH
Head to tail: male, 17–19 in. (43–50 cm); female, 16–18 in. (40–46 cm)

DISTINCTIVE FEATURES
Long red bill, the lower mandible longer than the upper, mandibles laterally flattened and knifelike; long wings, black above, white below; shortish, forked tail; short red legs; juvenile has boldly patterned brown-and-white upperparts

DIET
Surface-dwelling fish and crustaceans

BREEDING
Age at first breeding: 2 years; breeding season: March–September (North America); number of eggs: 2 to 4; incubation period: 21–30 days; fledging period: 35–42 days; breeding interval: 1 year

LIFE SPAN
Not known

HABITAT
Coastal lagoons, estuaries, beaches; also inland on rivers and lakes

DISTRIBUTION
East and south coasts of U.S.; Pacific and Caribbean coasts of Central America; much of eastern South America

STATUS
Fairly common to common

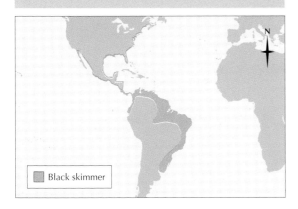

Black skimmer

the bill plowing into it. Showing astonishing aerobatic skill, the bill is maintained on a steady path through the surface of the water as the bird flies along. When the bill touches prey, it is snapped shut and the head nods downward and backwards, lifting the prey clear of the water. The birds usually skim the water at evening or in early morning, sometimes at night, and take small fish, crustaceans and other animals that live near the surface of the water. The three species of skimmers are the only birds in the world to use this specialized fishing technique.

When not feeding, black skimmers typically rest, preen or bathe. This bird indulges in a Florida lagoon.

Shrill calls

These sociable birds, which feed, nest and roost in flocks that may number several thousands, live on large rivers and lakes and along coasts, wherever there are sandbanks. Their distribution is regulated by the water level. At times of flood in tropical South America, the necessary riverine habitats disappear, and the black skimmers are confined to the coast. In North America, high river levels coincide with the breeding season, so the breeding grounds of North American black skimmers do not extend into rivers, but are found on barrier sand islands, salt marsh islands and sand banks around estuaries. When short of breeding or roosting space, skimmers will use grassland or even gravel roofs. In contrast, the nesting season coincides with the dry season in South America, when the water level is at its lowest. When roosting, all the members of a flock face into wind.

The calls of the black skimmer resemble a pack of hounds baying; the other two skimmers make shrill trills, screams or whistles. During the nesting season these calls are uttered when intruders approach their nests.

Cryptic chicks

Skimmers nest in loose colonies, with the nests scattered over a fairly wide area. The nests are no more than hollows in the sand with grooves where the sitting skimmer has rested its bill. Two to four buff, black-spotted eggs are incubated by both parents in turn, for 4½ weeks. The parents sit tight even when disturbed. The chicks leave the nest shortly after hatching. Their buff and brown down makes them very difficult to see as

they crouch in the sand, and they further conceal themselves by digging themselves into loose sand. If the chicks are threatened by predators, the parents perform a broken wing distraction display, diverting attention to themselves by suggesting they are injured and thus easy pickings for the predator. The chicks fly when they are about 6 weeks old.

Unique bill

When skimmer chicks hatch, the halves of their bills are equal in length and they can pick up food from the ground, but as their feathers start to grow, the lower half begins to get longer. By the time they are fully grown the bill has attained its unique shape. The bill can be opened very wide so that the upper half is clear of the water while the lower half is cutting through it. Near the base of the bill the edges are knifelike, allowing the skimmer to grasp slippery prey, and the neck muscles are very strong, enabling the skimmer to whip its prey out of the water as it speeds past.

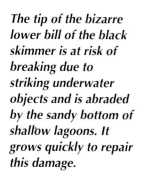

The tip of the bizarre lower bill of the black skimmer is at risk of breaking due to striking underwater objects and is abraded by the sandy bottom of shallow lagoons. It grows quickly to repair this damage.

SKINK

SKINKS HAVE NONE of the frills or decorations found in many other lizard families; most have an typical lizard shape with a rather long tail and very often limbs that are reduced or missing. These are adaptations to the burrowing way of life that is characteristic of skinks, and many spend most of their lives underground. Skinks are usually only a few inches (2–10 cm) long, the largest being the giant skink of the Solomon Islands, which is 2 feet (60 cm) long. The skink family contains over 1,300 species, found all over the warmer parts of the world. In some areas, such as the forests of Africa, they are the most abundant lizards.

Within the skink family there are all gradations from a running to a burrowing way of life. This can be illustrated by North American skinks. The Great Plains skink is an example of a running species, with a stocky body and relatively long legs. The many-lined skink, which is found in the United States east of a line from central Texas to Minnesota, is more elongated and has much shorter legs. The sand skink from central Florida is a truly burrowing lizard, whose forelegs are tiny, have only one toe, and fit into a groove on the side of the body. The hind legs are a bit longer, and each has two toes. There are no skinks in North America in which the limbs have been lost entirely, but lizards like this are found

elsewhere in the world: the burrowing skinks of southern Africa and Australia are examples.

Almost all skinks have a smooth, shiny skin. Burrowing species often have a protective, transparent spectacle covering the eye, and the ear openings are very small to stop them getting clogged up by soil or sand.

Skinks are found in a variety of habitats, both on the ground and beneath the surface, from the damp soil of forests to the sands of deserts. A few live in trees, but only the giant skink shows the obvious arboreal (tree-dwelling) adaptation of the prehensile (grasping) tail. Some skinks, such as the keel-bearing skinks, named after the projections on their scales, live on the banks of streams and dive into the water if alarmed. Some of the snake-eyed skinks live among rocks on the sea shore and feed on small crabs and marine worms.

Teeth to fit the diet

The main food of skinks is insects and other small animals, including young mice and birds' eggs. The insect-eating skinks have pointed teeth that crush their hard-bodied prey, and some of the skinks of Sri Lanka that feed on earthworms have backwardly curving teeth, which prevent the worms from escaping as they are being swallowed. Some of the species of burrowing skinks

A centralian blue-tongued lizard, Tiliqua multifasciata, *basks near Alice Springs in central Australia. Its colorful tongue is used in its gaping threat display.*

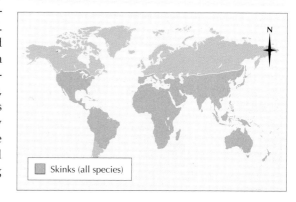

This sand skink, Scincus scincus, *in the dunes of Saudi Arabia, shows the large, fused head scales and the chisel-shaped head typical of sand swimmers.*

SKINKS

CLASS	**Reptilia**
ORDER	**Squamata**
SUBORDER	**Sauria**
FAMILY	**Scincidae**

GENUS AND SPECIES   **1,300 species, including giant skink,** *Corucea zebrata;* **five-lined skink,** *Eumeces fasciatus;* **many-lined skink,** *E. multivirgatus;* **Florida sand skink,** *Neoseps reynoldsi;* **Great Plains skink,** *E. obsoletus;* **Bermuda rock skink,** *E. longirostris;* **shingle-back,** *Trachydosaurus rugosus;* **and Kalahari burrowing skink,** *Typhlacontias gracilis*

ALTERNATIVE NAME
Shingleback: stump-tailed skink

LENGTH
80 percent of species: less than 15 in. (40 cm)

DISTINCTIVE FEATURES
Smooth, shiny skin; range from typical lizard-shaped to elongated and limbless

DIET
Almost all species: insects. Some medium-sized species: some plant material. Few large species: exclusively plants.

BREEDING
No generalizations possible. Half of skink species lay eggs, half produce living young.

LIFE SPAN
Large species: up to 20 years in captivity

HABITAT
Almost all terrestrial habitats

DISTRIBUTION
Worldwide except for northern Northern Hemisphere, southern Argentina and Chile, and many Caribbean islands

STATUS
Some species abundant. Some critically endangered: *E. longirostris, Lerista allanae.*

Skinks (all species)

feed almost exclusively on termites. The larger skinks are vegetarians and have broad, flat-topped teeth used for chewing.

Careful brooding

Some male skinks develop bright colors during the breeding season. When they meet, the males fight vigorously and may wound each other. A successful male follows a female, who allows him to catch her if she is ready to breed.

About half the skinks lay eggs; the others bear their young alive, some providing elaborate, placenta-like nutrition for the embryos while they are developing inside the mother. The porous eggs are usually laid in moist crevices so they do not dry out. Some skinks, such as the five-lined skink of North America, brood their eggs. The female curls around the eggs and stays with them until they hatch 4–6 weeks later, leaving them only to feed. She cares for them by defending them against predators and by maintaining them at the right moisture level, turning the eggs to prevent them from rotting.

Swimming in the sand

Some of the desert skinks, for example *Scincus mitranus* and *Chalcides ocellatus*, are called sandfish due to the way they swim through the sand. Their legs are well developed but they are held close into the body when moving. Propulsion comes from the flattened tail, which is reminiscent of the tails of amphibians or aquatic reptiles, such as the marine iguana. Another adaptation is a sharp, chisel-like snout that can cleave a way through the sand. Like other lizards, skinks have flexible skulls but their heads are strengthened for sand-swimming and burrowing by the fusing of the scales on the head.

SKIPPER

THESE BUTTERFLIES ARE characterized by their rapid, darting flight, which has given them their common name. They form a family of butterflies, the Hesperiidae, that are distinct from other butterflies. There is some doubt as to whether skippers should be called butterflies at all or whether they would be more correctly regarded as a group of moths. Unlike moths, skippers fly by day, have gradually widening antennae, regarded by some biologists as clubbed, and are generally more like butterflies than any of the Lepidoptera that are traditionally regarded as moths.

The antennae of a skipper almost always feature a slender extension beyond the clubbed region, which is usually hooked. The wing venation (arrangement of veins) is very basic and the body is thick and hairy. The skipper's head is wider than the thorax and the eyes are large and protruding.

Most of the skippers are small, but there are several genera with large wingspans. Two of these genera inhabit the American Tropics and subtropics, and another is found in eastern Asia. Of those found in the American Tropics and subtropics, the members of the genus *Pyrrhopyge* have extremely rapid and accurately controlled flight and are among the hardest of all butterflies to catch.

Sometimes included in a separate family, the Megathymidae, *Megathymus* is a genus of skippers confined to Mexico and the southern United States. In India and southeastern Asia some species of the *Gangara* and *Erionota* genera also have a 3-inch (7.5-cm) wingspan. They are unusual, as they fly at dusk and sometimes even come to artificial light after dark, as moths do.

Brown and tawny yellow are the most usual colors displayed by skippers, but some of the tropical species have bright metallic blues and greens. The species of *Argopteron*, found in South America, have the underside of the wings a solid metallic golden color.

The family contains in the region of 3,000 species, which are found worldwide, with the exception of Antarctica and New Zealand. Eight species are found in the British Isles, but most of have a limited distribution in the south. Only one, the dingy skipper, *Erynnis toges*, is found in Ireland. All the British species are common and widespread on the European continent, where a total of between 40 to 50 are known.

An Essex skipper, Thymelicus lineola. Skippers have features in common with both butterflies and moths.

2391

Silken cocoons

Skippers' larvae taper at both ends and are in most respects primitive in structure. They fasten together leaves with silk to form a shelter to live in, and pupate in rather flimsy silken cocoons or in rolled leaves fastened with silk. Many of the larvae feed on grasses. One of the large Asian species, *Erionota thrax*, lives on the leaves of the banana and is sometimes abundant enough to damage the plants. The larvae of the American genus of giant skippers, *Megathymus*, lives on yucca plants.

Moth or butterfly?

Although the distinction between moths and butterflies is largely artificial, one of the features that distinguishes them is the presence in most moths of a structure on the underside of the wings that joins together the fore- and hind wings. The component on the hind wing is called the frenulum and that on the forewing is called the retinaculum. These structures differ in male and female moths and may be one way in which scientists are able to identify a moth's gender.

In Australia skippers make up about one-third of the continent's butterfly fauna. One of the Australian species, called the regent skipper, *Euschemon rafflesia*, is unique among the Lepidoptera because the males have a well-developed frenulum and retinaculum and the females lack both. The regent skipper has metallic blue and white spots on the wings and is marked with bright red on the head and tail. In appearance, habits and life history this butterfly is clearly a skipper, and it illustrates the point that there is really no clear distinction between moths and butterflies.

The Lulworth skipper, Thymelicus octeon, *is rare in England but is common elsewhere in Europe. The pair below are mating.*

SKIPPERS

PHYLUM	**Arthropoda**
CLASS	**Insecta**
ORDER	**Lepidoptera**
SUPERFAMILY	**Hesperiidae**

GENUS AND SPECIES **Around 3,000 species, including Essex skipper, *Thymelicus lineola*; and great yucca skipper, *Megathymus yuccae***

LENGTH
Wingspan varies greatly between species. Great yucca skipper, *Megathymus yuccae*: 3⅛ in (8 cm); alpine skipper, *Oriesplanus munionga*: ¾–1 in (2–2.5 cm).

DISTINCTIVE FEATURES
Mothlike features: large head; antennae have gradually widening, hooked ends; stout body; generally regarded, however, as butterflies. Triangular forewings; hind wings may have tails.

DIET
Different species have different host plants; some eat smaller range of plants than others

BREEDING
Females usually deposit eggs singly on suitable host plants; number of generations per year differs between species

LIFE SPAN
Variable; some may survive for more than 1 year

HABITAT
Varies with species. Adults found around host plants in grasslands, scrub, wetlands, high-altitude areas, temperate forests, tropical forests

DISTRIBUTION
Most continents worldwide; not found in New Zealand

STATUS
Locally common

SKUNK

THE SKUNKS ARE MEMBERS of the family Mustelidae, along with the badgers, minks, otters and weasels. All mustelids have musk or stink glands at the base of the tail, but the skunk is perhaps the best-known mustelid and is able to squirt a nauseating fluid at its enemies.

The bold black-and-white color pattern of the skunk's fur makes it highly conspicuous and acts as a warning to any would-be predators. All 10 species of skunks have long fur and long, bushy or plumed black-and-white tails. The legs are short and the hind feet are plantigrade (both sole and heel are in contact with the ground when the animal walks). The soles of the feet are nearly naked; the five toes are webbed and each foot has strong, curved claws that are used for digging. The skunk has a small skull that is similar in shape to that of the Eurasian badger, *Meles meles*; there is only one molar in the top jaw and two in the bottom jaw.

Limited to the New World

The commonest skunk species is the striped skunk, *Mephitis mephitis*, ranging from southern Canada through most of the United States to northern Mexico. It grows up to about 30 inches (75 cm) long, including a tail that reaches about 12 inches (30 cm) in length. The females are usually smaller than the males. The fur, which is long and harsh with soft underfur, is black with white on the face and neck, dividing into two white stripes diagonally along the sides of the body. The hooded skunk, *M. macroura*, which is native to the southwestern United States and Central America, is similar to the striped skunk but the tail is longer than the head and body and the white stripes are more widely separated. The three species of spotted skunks, *Spilogale pygmaea*, *S. gracilis* and *S. putorius*, range from British Columbia through most of the United States and Mexico into Central America. They are distinguished by their small size, being only 22 inches (55 cm) maximum length, of which 9 inches (22.5 cm) is accounted for by the tail. They also have a white spot on the forehead and a pattern of white stripes and spots.

The five species of hog-nosed skunks, *Conepatus mesoleucus*, *C. leuconotus*, *C. semistriatus*, *C. chinga* and *C. humboldti*, are not as well known as the other species of skunks. They are

Skunks are able to adapt to a wide range of habitats, provided there is a food source nearby. However, they are most commonly found near woods.

The striped skunk has two distinctive white stripes running down its sides. Its bold black-and-white marking is intended to dissuade would-be predators.

SKUNKS

CLASS	**Mammalia**
ORDER	**Carnivora**
FAMILY	**Mustelidae**

GENUS AND SPECIES **Striped skunk, *Mephitis mephitis*; hooded skunk, *M. macroura*; spotted skunks, *Spilogale pygmaea, S. gracilis* and *S. putorius*; hog-nosed skunks, *Conepatus mesoleucus, C. leuconotus, C. semistriatus, C. chinga* and *C. humboldti***

WEIGHT
7–14 oz. (200–400 g)

LENGTH
Head and body: 4⅓–19⅓ in. (11–49 cm); tail: 2¾–17⅓ in. (7–44 cm)

DISTINCTIVE FEATURES
White markings on black coat, spotted with lines in spotted species, small white patches or distinct white lines from head to tail in others; small head; short legs; long tail

DIET
Small rodents and birds; insects; vegetation

BREEDING
Striped skunk. Age at first breeding: 1 year; breeding season: February–April; number of young: usually 5; gestation period: about 65 days; breeding interval: usually 1 year.

LIFE SPAN
Up to 5 years; more than 12 years in captivity

HABITAT
Varied, including scrub, woodland, desert and rocky habitats

DISTRIBUTION
Mainly North and Central America; some hog-nosed skunks also in South America

STATUS
Striped skunk and several other species: common. *S. pygmaea*: rare; subspecies *C. mesoleucus telmalestes*: probably extinct.

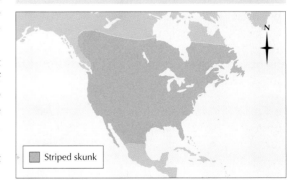

Striped skunk

found in the southwestern United States and are the only genus of skunks with representatives in South America. They are much the same size as the striped skunk but with shorter, coarse hair; the top of the head, back and tail is usually white. They have a long, naked, piglike snout that they use for rooting up insects.

Noxious spray

Skunks are found in a variety of habitats, including woods, plains and desert areas, although they avoid wetlands and dense forest. They live in burrows that they dig for themselves or in the abandoned burrows of badgers, foxes or woodchucks, or under buildings, denning up by day and only coming out at night to forage. Although they are not true hibernators, most skunks, especially those in the northern parts of their range, settle down in their dens and sleep for long periods during the cold weather. The dens are lined with dry leaves and grass. Occasionally several skunks den together. Spotted skunks, however, are active throughout the year.

When it is disturbed or attacked, a skunk lowers its head, erects its tail and stamps a warning with its front paws. It may also chirp, growl, bark or purr by way of warning. If the intruder remains undeterred, the skunk turns its back and squirts an amber-colored, foul-smelling fluid, composed of a chemical called mercaptan, from glands situated on either side of the anus. It is able to project the fluid for as far as 12 feet (3.6 m) with unerring accuracy and can repeat the spray seven or eight times if necessary. This pungent spray can cause temporary blindness if it touches the eyes and its odor can be detected half a mile (0.8 km) away. It is highly potent and extremely difficult to remove. A striped skunk turns its back to a predator and eject its fluid at it, while a spotted skunk may discharge from a handstand position.

Skunks have very poor eyesight. They can only see well up to about 3 feet (90 cm). This may explain why the animal is ready to act so defensively at the first hint of a predator.

Carnivorous feeder

Skunks feed mainly on vegetation and insects such as beetles, crickets, grasshoppers and caterpillars. They also take mice, frogs, eggs, small birds and crayfish, and occasionally they cause damage by entering poultry runs, killing the birds and taking the eggs. As the winter approaches, the striped skunk becomes fat, adding leaves, grain, nuts and carrion to its diet. Some skunks feed on snakes. The hog-nosed skunks in the Andes are immune to the venom of pit vipers, and spotted skunks may be resistant to rattlesnake venom.

Blind and hairless kits

The striped skunk breeds between February and April, and the mating season is preceded by boisterous but relatively harmless fighting among the males. After a gestation of 60–77 days, 4 or 5 young, occasionally up to 10, are born in the den. The young skunk, or kit, weighs about an ounce (28.3 g) at birth and is blind, hairless and toothless, but the black-and-white pattern shows plainly on its skin. Its eyes open in 21 days and it is weaned in 6–7 weeks. Toward the end of this period it is taken on hunting trips by its mother. By the fall the family breaks up and each youngster goes its own way to fend for itself. Skunks become mature in about a year and have lived for 10–12 years in captivity.

The young of spotted skunks are usually born in late May or June although they may be born at any time in the southern parts of their range, where two litters may be raised in a year. As in the striped skunk, the usual litter size is four or five, but the young weigh only ⅓ ounce

(9.4 g) at birth and are only 4 inches (10 cm) long. They are not weaned until they are 8 weeks old, when they begin to take insects.

Left alone by most predators

Most predators give skunks a wide berth, but pumas and bobcats occasionally kill and eat them when their usual prey is scarce. Only the great horned owl preys regularly on the skunk, and it generally bears the noticeable odor of skunk about it. Many skunks are killed every year on the road by cars, particularly at dusk. They appear not to have learned to run away from a car and instead sometimes stand their ground and eject their fluid in futile defiance, as they would at a living predator.

Some skunks are still trapped for their fur, and in South America the hog-nosed skunk is hunted by the local people, as its meat is said to have curative properties. The skin is used for capes and blankets.

This striped skunk has assumed a defensive posture from which it can emit its potent musk at an intruder.

SKYLARK

A skylark searching for seeds on a beach in winter. The skylark's summer diet consists mainly of small invertebrates.

LARKS ARE FAMOUS FOR their songs, but few are as well known as the skylark, which has inspired poets, particularly Percy Bysshe Shelley. A 19th-century English Romantic poet, Shelley referred to the bird as "blithe spirit." Although its song is familiar to many people who live in its range, the skylark itself is often overlooked.

The skylark, *Alauda arvensis*, is a drab brown bird, about 7–7½ inches (18–19 cm) long. The plumage is brown streaked with black on the upperparts, head and breast and whitish underneath. The skylark could easily be dismissed as just another small, brown bird if it were not for the white feathers edging the tail and the short crest. The latter is much less conspicuous than that of some other larks, such as the crested lark (discussed elsewhere). Sometimes it appears as if a skylark has no more than a ruffled head and often the crest is laid flat so that it cannot be seen.

Skylarks breed in most of Europe, except the extreme north of Scandinavia, in North Africa and across Asia to Japan and the Bering Strait. Because of the popularity of its song, the skylark has been introduced into many other parts of the world, such as New Zealand, Hawaii and Vancouver Island off the western coast of Canada. A second species of skylark, the small or oriental skylark, *A. gulgula*, breeds in southern Asia east as far as the Philippines. The small skylark fills a similar ecological niche to the northern skylark, living in natural grasslands at a fairly wide range of altitudes and avoiding wooded terrain.

Bird of open country

Over most of Europe and North Africa the range of the skylark overlaps that of the woodlark, *Lullula arborea*. The latter species is smaller and has conspicuous white eye stripes. Its song is different from that of a skylark, and during the nesting season the woodlark is found on the edges of woods or among scattered trees. The skylark is typically found on pastures, steppes, waste ground, hillsides, moors and marshes, but not near trees.

In the fall skylarks migrate south in flocks, but to the casual observer it may appear as if there has been no migration at all because skylarks, in temperate Europe at least, are as common in winter as they are in summer. Shortly after breeding, as skylarks fly south, more skylarks arrive from the north to take their place, some just passing through but others staying for the winter. In the British Isles the native skylarks leave in September–November, and during this time they are replaced by birds from northern and central Europe. Those from the north stay, living in flocks, whereas those from central Europe continue their journey southward.

Skylarks feed on a mixture of animal and vegetable food. Over half their food is plants, particularly seeds of corn, chickweed and the leaves of clover. They also eat a variety of small animals, such as insects, millipedes and spiders.

Long-playing song

The song of the skylark can be heard nearly all year-round; even the winter visitors sing, and skylarks also sing while on migration. The song is a clear rather tuneless warble that may continue unbroken for up to 5 minutes, which poses the problem of how the skylark can breathe at the same time. The song is delivered during a song flight, similar to that of the pipits. The skylark flies up vertically, starting to sing almost as soon as it clears the ground. The song continues until the skylark is little more than a speck in the sky poised on fluttering wings. The skylark then sinks as gradually as it ascended, and finally it drops to the ground and disappears in the grass. The song is sometimes delivered from a post or from the ground.

The small nest, just 2½ inches (6 cm) across, is always built on the ground, usually among grass. It is very difficult to find because it is made of

SKYLARK

CLASS	**Aves**
ORDER	**Passeriformes**
FAMILY	**Alaudidae**
GENUS AND SPECIES	***Alauda arvensis***

ALTERNATIVE NAME
Northern skylark

WEIGHT
1–1¾ oz. (28–50 g)

LENGTH
**Head to tail: 7–7½ in. (18–19 cm);
wingspan: 12–14 in. (30–36 cm)**

DISTINCTIVE FEATURES
**Small size; short crest; proportionately
long legs with large feet and long claws;
inconspicuous brown, buff and white
plumage with dark streaking on upperparts,
head and breast; white outer tail feathers**

DIET
**Summer: insects, spiders and millipedes;
autumn and winter: grain, seeds and leaves**

BREEDING
**Age at first breeding: 1 year; breeding
season: eggs laid late March–August;
number of eggs: usually 3 to 5; incubation
peiod: 11 days; fledging period: 18–20 days;
breeding interval: up to 4 broods per year**

LIFE SPAN
Up to 8 years

HABITAT
**Open habitats with short grass, including
fields, meadows, heaths, steppes, hillsides,
salt marshes and waste ground**

DISTRIBUTION
**Most of Europe, east through Central Asia
and southern Siberia to Japan, south to
North Africa and Middle East**

STATUS
**Very common, but numbers declining on
intensively farmed agricultural land**

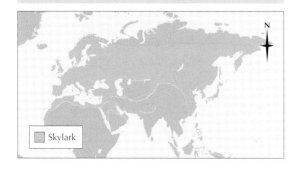

Skylark

grass stems and is hidden by additional stems that form a tangle over it. There is, however, a pathway leading through the grass to the nest, and sometimes a skylark can be seen landing and running to its nest.

There are usually three or four eggs in a clutch, greenish white with brown speckles. They are incubated by the female alone for 11 days. The young are fed by both parents and leave the nest when just over 1 week old but do not fly for almost 3 weeks.

Almost always singing

The skylark is one of the few birds that sings almost all year-round. The time spent singing depends on a variety of circumstances but seems to have neither rhyme nor reason. For example, the male's singing reaches a peak in March or April—understandable because he is then holding a territory and courting. It becomes relatively infrequent during nest-building and incubation, although he does not take part in these activities, and there is no obvious reason why he should not be singing. Then he sings again while the chicks are being fed even though he helps in the feeding. The one thing that does seem to upset a skylark's song is fog or very high winds. Skylarks sing less under these weather conditions, although some of them still continue to sing from the ground.

Skylarks are famous for their beautiful warbling song, which is usually delivered high in the air but is also performed on fence posts or even on the ground.

SLATER

Many sea-slaters are nocturnal scavengers that hide under rocks and in crevices during the daylight hours.

SLATERS AND WOOD LICE ARE part of the same group of animals, the Isopoda. However the name slater tends to be applied more often to a species related to wood lice living by the sea. Wood lice are described separately in this encyclopedia, but the sea-slater merits special treatment not so much for its unusual size, up to 1½ inches (3.75 cm) in males and 2 inches (5 cm) in females, but for its half-terrestrial, half-marine way of life. Sea-slaters are drably colored: blackish, dark grayish green or light dirty brown. There are seven free thoracic segments, each known as a pereonite, and together they form a structure called the pereon. The first thoracic segment is fused to the head, the fourth segment being the widest part of the roughly oval body. Behind the thorax is an abdomen of five free segments called pleonites, with the sixth segment being fused to the telson (terminal segment) to form a structure called the pleotelson. The abdomen has respiratory lobes on its appendages and the telson bears two pairs of sensory tails. On the head are two large, black, unstalked compound eyes and two pairs of antennae, the first tiny and the second very long.

Slippery slaters

The sea-slater, *Ligia oceanica*, lives on the coasts of Europe and North America, and there are several other species on the European and American coasts. Other species are found on coasts throughout the world. Although they do sometimes venture farther down the beach, sea-slaters

live mostly just above high tide level, hiding by day in cracks and crevices, so that their abundance is not often appreciated. They are not found far inland, only within the zone splashed by sea spray. However, in some places with steep cliffs, they can live 450 feet (135 m) up. Slaters also occur up the sides of estuaries and in salt marshes. Their abundance on quays has earned them the alternative names of quay lice, quay lowders or, in North America, shore-roaches. At certain times on the rocky shores of Japan, slaters are so abundant that fishermen sweep them up with brooms for bait. Slaters are fast runners, each leg making over 16 steps per second, a rate that is equaled by few other arthropods. With a step of about ⅗ inch (1.5 cm), a large slater scuttles at slightly more than 1 mile per hour (1.6 km/h), very fast for such a small animal. Slaters prefer darkness; even moonlight does not usually tempt them to emerge.

Sweating in the sun

On a dark night when the tide is out, slaters descend the shore to feed. They are particularly fond of bladder wrack and other seaweeds, but they are general scavengers of organic matter, and in captivity at least they may eat each other. Because they hunt at night, they avoid being preyed on by birds, but many are caught by crabs. Drying up is another problem, because slaters readily lose water by evaporation through their cuticle. Nevertheless, they are fairly tolerant of water loss. Their large size helps in this respect and can sometimes use evaporation to combat yet another problem, that of overheating as the sun beats down on the shore. Scientists have discovered that in dry air, when evaporation from the body is critical, the internal temperature of a slater may be as much as 44° F (24° C) below that of the surrounding air. The slater loses heat by evaporation, in the same way that humans do when they sweat. Therefore it is sometimes beneficial for a slater to move out from under a stone and into sunlight, where more evaporation can take place.

The widespread species *L. oceanica* may be found covered by the tide and has been known to survive months in seawater in the laboratory and to fast during this time. It has also survived 1½–8 days in distilled water. Several species, for example, *L. baudiensis* of Bermuda, which follows the ebbing tide and retreats as the sun dries the beach, cannot stand long immersion and, if pursued, enter the water only if there is no other escape route.

Young born in a pouch

The mature female has a thin plate on each of the first five pairs of legs that curves down and inward and overlaps the opposite and adjacent plates. Together they make up a brood chamber. The male fertilizes the eggs after they are laid in this chamber, using a pair of grooved and two-jointed styles on his second pair of abdominal appendages to insert sperm into the brood chamber. After the eggs have hatched, a brood of about 80 young, known as manca, are sheltered

in the pouch, then released into the sea. They look like miniatures of their parents. The young disperse primarily by crawling. Females with broods can be found throughout the year, but particularly in spring. In warmer parts of the world, plagues of slaters frequently occur. When this happens, the slaters often invade houses near quays and wharfs.

As in other crustaceans, growth in slaters involves periodic casting of the skin. Some days before a molt the underneath of the first four segments of the thorax begins to develop a chalky whiteness. The upper surface of these segments also becomes lighter in color. The old skin then splits behind these segments and the hind portion is shed. The front part is not cast until about four days later. As a result of this two-stage molt, some slaters often have their hind end lighter than the front. The change in size after molting is not very obvious. Sea-slaters probably live about three years.

Rhythmic color change

With their drab colors, sea-slaters tend to merge with their surroundings even when not actually hidden away. Their ability to do this is helped by a tendency to grow lighter or darker according to the darkness of the background, by the contraction or expansion of their chromatophores (pigment-bearing cells). The stimulus to do this is through the eyes; a blind slater does not respond to a change in background. There is, however, another response that occurs even if the slater is blind: a blanching in darkness and a darkening in the light. These responses involve two hormones with opposing actions on the color.

With their segmented bodies and jointed appendages, sea-slaters are typical examples of the class Crustacea.

SLATERS

PHYLUM	**Arthropoda**
CLASS	**Crustacea**
ORDER	**Isopoda**
FAMILY	**Several, including *Idoteidae*, *Sphaeromatidae* and *Cirolanidae***
GENUS AND SPECIES	**About 5,000 species**

ALTERNATIVE NAMES
Shore-roach; quay louse; quay lowder

LENGTH
Varies greatly. *Ligia oceanicus*: up to 1⅛ in. (3 cm).

DISTINCTIVE FEATURES
Similar to wood lice; gray body; broadly oval shape; 2 antennae; 2 nonstalked eyes; head fused to first thoracic segment; sixth segment fused to telson

DIET
Varies with species; slaters are scavengers, parasites and opportunists

BREEDING
Young (manca) develop in female's brood pouch; emerge crawling as miniature adults

LIFE SPAN
Not known, but may be several years

HABITAT
Varies considerably: intertidal (*Ligia*), fresh water, dry land or aquatic

DISTRIBUTION
Worldwide

STATUS
Locally common

SLOTH

A sloth's fur grows from its belly to its back; the reverse is true of other mammals. Pictured is the two-toed sloth, Choloepus didactylus.

SLOTHS ARE TREE-LIVING mammals of South America and southern Central America that spend nearly all their lives hanging upside down. They belong to the suborder Xenarthra of the order Edentata, which means "toothless." Edentates also include anteaters and armadillos, although the anteaters are the only members of this order that wholly lack teeth. Sloths have teeth in the cheeks; there are nine on each side and they grow throughout life.

There are five species of sloths, grouped into the two genera *Choloepus* and *Bradypus*. Their bodies show some remarkable adaptations for an upside-down life in the trees. Sloths hang by means of long, curved claws, similar to meat hooks, and their hands and feet have lost all other functions, the fingers and toes being united in a common fold of skin. The arms are longer than the legs and the pelvis is small. The back muscles, which are well developed in other animals, are weak in sloths. The head can turn through 270 degrees, so that it can be held almost the right way up while the rest of the body is hanging upside down.

Sloths have short, fine underfur beneath an overcoat of longer, coarser hairs. The hair lies in the opposite direction to that of other mammals, from belly to back, thus allowing rainwater to run off it. The individual hairs are grooved and are usually infested by single-celled algae, which give sloths a greenish appearance. All sloths have inconspicuous ears, especially those belonging to the genus *Choloepus*.

The five sloth species are divided into two-toed and three-toed sloths (though, strictly speaking, they all have three toes on each hind foot; the difference lies in the foreclaws, or fingers.) Three-toed sloths, genus *Bradypus*, grow to about 27½ inches (70 cm) long with a stumplike tail of about 3½ inches (9 cm). Two-toed sloths, *Choloepus*, are a little larger, at up to 29 inches (74 cm), but lack a tail. Sloths live in forests, the three-toed sloths from Honduras to northern Argentina and the two-toed sloths from Venezuela to Brazil.

Slow-motion animals

As is common for nocturnal animals of the tropical American forests, little is known about sloths' habits. Their movements along the branches are so slow that they were once erroneously thought to spend their whole life in one tree. They eat, sleep, mate, give birth and nurse their young upside down, although they do not hang from branches all the time, and occasionally sit in the fork of a tree. Sloths sleep with the head on the chest between the arms.

Sloths come to the ground only occasionally. They may do so weekly, in order to urinate and to defecate, acts that they usually perform in set locations against the bases of trees. It is likely that they also descend to the ground in order to reach another tree when they are not able to

SLOTHS

CLASS	**Mammalia**
ORDER	**Xenartha**
FAMILY	**Megalonychidae; Bradypodidae**
GENUS AND SPECIES	**Two-toed sloths, *Choloepus didactylus* and *C. hoffmani*; three-toed sloths, *Bradypus variegatus*, *B. tridactylus* and *B. torquatus***

WEIGHT
***Choloepus*: 8¾–18¾ lb. (4–8.5 kg); *Bradypus*: 4½–13¼ lb. (2–6 kg)**

LENGTH
***Choloepus*: head and body: 21–29 in. (54–74 cm). *Bradypus*: head and body: 16¼–27½ in. (41–70 cm); tail: ⅘–3½ in. (2–9 cm)**

DISTINCTIVE FEATURES
Dense fur. 3 curved claws on hind limbs. *Bradypus*: forelimbs much longer than hind limbs, end in 3 claws. *Choloepus*: forelimbs slightly longer than hind limbs, end in 2 claws

DIET
Leaves, fruits, bark, twigs; may absorb nutrients from algae through skin, and by licking

BREEDING
Age at first breeding: 3–5 years; number of young: 1; gestation period: 330–360 days (*Choloepus*), 150–180 days (*Bradypus*); breeding interval: 14–16 months (*Choloepus*), 1 year (*Bradypus*)

LIFE SPAN
Up to 20 years

HABITAT
Dry and tropical forest

DISTRIBUTION
Central America and South America south to Argentina

STATUS
Locally common, but numbers decreasing due to habitat loss. *B. torquatus*: endangered.

Sloths | Three-toed only | Three-toed and two-toed

travel overhead by using branches and creepers. They are just about able to stand on their feet, but cannot walk on them and move by sprawling on their bellies and dragging themselves forward with their hands. By contrast, they swim well.

Defense and feeding

Sloths can defend themselves well by slashing with their claws and biting. It is possible that this is sufficient defense against their main predators, jaguars and ocelots, but it is difficult to imagine a nimble cat being unable to outmaneuver a sloth. They may gain greater protection from the camouflage that the green algae afford them and by their limited movements, which give them a passing resemblance to a mass of dead leaves.

Sloths eat mainly leaves, shoots, bark and some fruit, which they may hook toward their mouths with their claws. Their stomachs are

A sloth (Bradypus tridactylus, above) uses its long, strong claws to support itself. The animal leads a primarily arboreal lifestyle, eating, sleeping, mating and giving birth in trees.

complex, like those of ruminants such as cattle and sheep, and contain cellulose-digesting bacteria. A full stomach may account for up to one-third of a sloth's total body weight and it may take a month for a meal to be digested.

All sloths are able to maintain low, though variable, body temperatures of between 86–93° F (30–34° C). These levels fall during periods of inactivity, wet weather and at night, when the surrounding atmospheric temperature falls. It is important for sloths to vary their body temperature, as their metabolic rate is less than half that which would be expected for animals of their size. They also have reduced muscles and cannot afford to lose energy by shivering. Sloths frequent the exposed crowns of trees and vary their body temperature by moving in and out of the warm sunshine.

The greenish tinge to a sloth's fur is caused by algae that infest the animal's hairs.

A solitary existence
Most adult sloths are solitary animals and there is still considerable scientific debate about their methods of communication. Males probably make their presence known by smearing secretions from an anal gland onto tree branches. It is also possible that dung piles serve as meeting places for males and females.

Sloths produce a limited range of calls. Two-toed sloths hiss when they are disturbed, while three-toed sloths utter piercing *ai-ai* whistles through their nostrils.

One young
Sloths breed throughout the year. A single young is born at the beginning of the dry season after a gestation of 150–180 days in those species belonging to the genus *Choloepus*. By contrast, in *Bradypus* species, the gestation period may last up to 360 days. The newborn sloth immediately hooks itself into the fur of its mother's breast, where it remains until it is old enough to leave. This usually takes 6–9 months, during which the young feeds on any leaves it is able to reach.

After the young is weaned, it inherits part of the mother's range, as well as the mother's taste in the leaves of particular trees. This preference for a specific range of leaves enables several sloths to share the same home range without competing with each other for food. Sloths have lived for up to 30 years in captivity.

Algal lodger
Sloths may absorb nutrients from the green algae that infest their fur. However, their fur harbors another life form, a moth similar to a clothes moth.

The three species of pyralid moths have been found on sloths belonging to both the *Choloepus* and *Bradypus* genera. They are about ⅓ inch (8.5 mm) long with flattened bodies and are able to run agilely through the dense mat of the sloth's hair. This makes the moths difficult to collect, particularly as the collector has to avoid the sloth's attempts to defend itself in order to reach them.

The reason why moths live in sloths' hair has not yet been established. The moths do not feed there, nor have their eggs or caterpillars been found in the fur. In 1976 two American zoologists discovered that the larvae feed on the sloth's dung.

SLOTH BEAR

ALSO KNOWN AS the Indian bear, the sloth bear does not look like a typical bear because of its long, shaggy hair and thick, loose lips. Its feet have well-developed, curved claws that grow up to 3 inches (7.5 cm) long and with which the bear is able to suspend itself upside down, in the manner of a sloth. The coat is predominantly black or blackish brown, occasionally reddish, and the hair is particularly long on the back of the neck and between the shoulders. The long muzzle is dirty white or gray and there is a characteristic white, sometimes brown, U- or V-shaped mark on the chest.

Sloth bears grow to about 6 feet (1.8 m) long, of which 4–5 inches (10–12.5 cm) is tail, and stand 26–32 inches (65–80 cm) at the shoulder. Some individuals may weigh up to 330 pounds (150 kg); females are smaller than the males but have denser fur.

Lowland bears

Sloth bears live in Sri Lanka and in India from the south to the Himalayan foothills. In Sri Lanka they are found in the low country of the dry zone, and in India they live in lowland jungles and are rarely found at any great altitude. Sloth bears are active mainly at dusk and dawn and often at night. They keep well away from human settlements, although they frequently raid crops and attack people, and they sleep in the cover of low vegetation or among rocks. Their usual gait is a slow, shambling walk, but when necessary sloth bears can break into a gallop and move faster than a running human. (Most bears can run up to 30 miles per hour [48 km/h].)

A taste for ants and termites

Sloth bears are omnivorous and their wide-ranging diet includes insects, specially termites and their larvae, sugarcane, carrion, birds' eggs, fruit, flowers and roots. They climb well, scaling trees to rob the nests of bees and birds; their long claws are very useful for foraging in trees for fruit and flowers. The diet varies with the season. In the wet season, when the ground is soft, termite hills are easy to break open and there are plenty of small animals to be found under fallen logs or in leaf litter. When the ground is baked hard and small animals are difficult to find, fruit forms a larger part of the bears' diet.

Several mammals have taken to feeding on ants and termites, including another carnivore, the aardwolf (discussed elsewhere). The sloth bear catches termites using a very different

Although the sloth bear is the most common bear in India, habitat destruction and poaching have severely reduced its numbers in recent times.

The sloth bear is distinguishable by the dramatic white chest marking. The specimen above was photographed in a zoo in Miami, Florida.

SLOTH BEAR

CLASS	**Mammalia**
ORDER	**Carnivora**
FAMILY	**Ursidae**
GENUS AND SPECIES	***Melursus ursinus***

ALTERNATIVE NAME
Bhalu (Hindi)

WEIGHT
275–330 lb. (125–150 kg)

LENGTH
Head and body: 56–68 in. (140–170 cm); shoulder height: 26–32 in. (65–80 cm); tail: 4–5 in. (10–12.5 cm)

DISTINCTIVE FEATURES
Shaggy, glossy black fur; white U- or V-shape on chest; grayish yellow tip to muzzle and feet; elongate muzzle; very developed claws, particularly on forelimbs

DIET
Fruits, wild honey and tree sap; insects, particularly termites; carrion

BREEDING
Age at first breeding: 3–4 years; breeding season: May–July (India), longer in Sri Lanka; number of young: 1 or 2; gestation period: 6–7 months; breeding interval: 2–3 years

LIFE SPAN
25–40 years in captivity

HABITAT
Forest, including mixed savanna forest, and forest edge

DISTRIBUTION
India, Bhutan, Nepal and Sri Lanka

STATUS
Vulnerable due to trade in bear gall bladder for medicine; estimated population: about 10,000

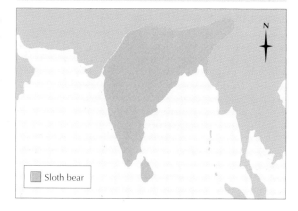

method from that employed by other ant-eating mammals. Common features of these animals are a long snout, a long tongue and teeth that are either small or missing altogether. After they have broken open a termite hill, they literally wipe up the termites with their sticky tongues. By contrast, the sloth bear sucks up its termite prey, rather in the manner of a vacuum cleaner. The muzzle is adapted for this purpose: the nostrils can be closed, the inner upper pair of incisors are missing and the loose lips can be formed into a tube with which to suck up prey. When a sloth bear has broken into a termite's nest, it immediately inserts its head and blows violently, in order to drive away dust and debris, after which it sucks up both the termites and their grubs. The sloth bear's sucking and blowing can be heard up to 200 yards (180 m) away.

Carried by mother

Sloth bears mate year-round in Sri Lanka but the mating season appears to be confined to June in India. Scientists believe that they have only one mate. Courtship is boisterous, but after mating the male is driven away. Usually two cubs, sometimes one and rarely three, are born 7 months later in a den among boulders or in a cave. When they are 2–3 months old they leave the den and accompany their mother. They sometimes ride on her back, clutching the long hair between her shoulders. The cubs stay with their mother for 2–3 years.

Dangerous when scared

The sloth bear is regarded as one of the most dangerous animals in the jungles of India and Sri Lanka, and villagers in Sri Lanka are said to fear it more than any animal except a rogue elephant. However, attacks are usually due to the sloth bear being frightened by the sudden appearance of a human rather than because of any inherent aggression on the animal's part. The bear's vision and hearing are poor, and if it is approached from downwind it may not notice a human's presence until the last moment. The sloth bear may then easily become panicked at the sudden discovery of a newcomer and charge, knocking over and mauling its victim. Habitual rogue bears have been known to attack people.

The sloth bear's reputation as a dangerous animal almost certainly explains why it was not studied scientifically for some time. This ignorance also explains how the sloth bear acquired its name. Sloth bear skins were sent to Europe by big-game hunters toward the end of the 18th century. The notes accompanying the skins described an animal with a trunklike snout, which swung acrobatically through the branches in the jungles of India and Sri Lanka and uttered a cry similar to that of a child. The first scientists to examine the skins referred to them as "the nameless animal." Influenced by the descriptions of the animal swinging through the trees, George Shaw of the British Museum classified the animal among the sloths. In 1810, when a live sloth bear was brought to Paris, it became clear for the first time that it was not related to the sloth. However, the name sloth bear persisted.

The sloth bear is the most common bear in India. However, a number of factors have brought about a decline in its population. Deforestation has reduced the bear's range, while poaching has also had an adverse effect on numbers. Sloth bear gall bladders are imported into Japan to be used in traditional medicine, and although the bear is classified as endangered by the Indian government, and it is now illegal to import bear parts from India, the trade persists.

Few measures exist to protect the sloth bear, and those laws that exist to prevent its being hunted are either ineffective or not enforced. The future of the sloth bear remains precarious.

The sloth bear is perhaps more special-ized to suit its dietary needs than any other bear. It can close up its nostrils and form its lips into a tube with which to suck up its insect prey.

SLOWWORM

counties of England. The color, which may vary from light blue to deep ultramarine, may be present in spots or very occasionally in stripes, sometimes so closely set that the animal appears blue all over. Only the male blue-spotted slowworms bear these blue markings.

Not fussy

Slowworms inhabit a remarkably wide range of habitats. They are found in gardens, sometimes in large, densely-populated cities, and in hedgerows, orchards, open woodland and scrub, on sea cliffs and in mountains. They are especially abundant on waste ground and are one of the animals that, in England, can frequently be found in churchyards. They are very common animals in many places, but are rarely seen. They spend a lot of time buried in leaf litter and soil. They are mostly active at dawn and dusk, because the earthworms and small grey slugs on which they predominantly feed are also most active then. Sometimes individuals are seen during daylight hours, especially when the sun is shining. In the spring males use the warmth to speed the development of the reproductive system, while in the summer females use it to speed the growth of their developing embryos.

Basking in secret

Slowworms have another means of warming up: they creep into the spaces beneath flat, thin stones lying on the ground. If the sun is shining, the surface warms up, and some of the heat is conducted to the underside. Some of this heat can be used to warm a slowworm lying beneath. Even on overcast days, there is sometimes enough heat radiation to enable lizards to gain a slightly higher body temperature than their surroundings. Using this technique, they can receive the advantage of warmth without exposing themselves to predators by basking. In recent times, slowworms have discovered that trash, such as pieces of corrugated iron sheeting, can often act more effectively as sources of warmth. Biologists have taken advantage of this habit: to estimate the population of slowworms in an area, previously a very difficult task, they can leave flat pieces of metal and wood on the ground for a few days, then count the slowworms underneath.

Neither a worm nor a snake, the slowworm is actually a lizard, related to the glass lizard, Ophisaurus attenuatus, of North America. Mating in slowworms can be a violent procedure.

THE SNAKELIKE SLOWWORM, occasionally known as the blind worm, is in fact a legless lizard. Internally there are vestigial shoulder and hip girdles, evidence that its ancestors once moved on four legs. A slowworm has eyelids like other lizards. The two halves of its lower jaw are joined in front, another lizard characteristic, and its tongue is notched, not forked like that of a snake. An average large slowworm is around 1 foot (30 cm), but the biggest slowworm ever recorded was a female, 20.6 inches (52 cm) long.

The head of a slowworm is small and short, not so broad as the body immediately behind it and larger in the male than in the female. Fully grown males are more or less uniform in color above and on the flanks. They may be light or dark brown, gray, chestnut, bronze or brick red, and one variety is even copper colored. The belly usually has a dark mottling of blackish or dark gray. The female often has a thin dark line down the center of the back and another on the upper part of each flank, and her belly is usually black.

The slowworm is found throughout Europe, including Britain, and eastward to the Caucasus and European Russia. In Sweden it extends as far as latitude 65° N. A variety of the slowworm, known as the blue-spotted slowworm, is widely distributed over Europe, including the southern

Not a worm, and not always slow

Although the word *worm* has no formal zoological definition, zoologists generally agree that all worms should be classed as invertebrates. Slowworms are clearly lizards and not worms. When they are hunting, they usually move in a sedate and deliberate-looking manner, but if an individual slowworm is disturbed, it can slither away remarkably fast if it has been warming itself beforehand.

Ovoviviparous female

Adult slowworms of both sexes usually bear many scars, originating in several ways. Mating is from late April to June, when there is a great deal of fighting between the males, each trying to seize the other by the head or neck. Once a hold has been obtained, the males writhe and roll about together. Mating also involves biting. In this case, the male grasps the female's neck in his jaws and twines his body around hers. Cats are especially ferocious hunters of slowworms; if an individual gets away, it is frequently damaged by the encounter.

Female slowworms do not lay eggs, but retain the eggs in the form of thin, membranous envelopes within the uterus until they are ready to hatch. Many lizards do this, especially in cool climates; the process is known as ovoviviparity. Many female slowworms produce a litter only every two years, the young being born in August or September. The young slowworms are 3–3½ inches (7.5–9 cm) long when born, and quite unlike their parents, with a black stripe along the back and a black underside. Very active, they are able to fend for themselves from the moment of birth, catching insects, but showing a marked preference for any slugs small enough to eat.

Gardener's friend

Because slowworms feed predominantly on slugs, so often a pest on tender green vegetables, increasing numbers of gardeners encourage them. However, slowworms are still killed in large numbers, out of the mistaken fear that they are snakes. In fact, slowworms are harmless and beneficial to humans.

SLOWWORM

CLASS	**Reptilia**
ORDER	**Squamata**
SUBORDER	**Sauria**
FAMILY	**Anguidae**
GENUS AND SPECIES	***Anguis fragilis***

ALTERNATIVE NAME
Blind worm (increasingly rarely used)

LENGTH
Up to 20½ in. (52 cm)

DISTINCTIVE FEATURES
Snakelike form; eyelids and external ear openings; smooth, shiny skin; few markings

DIET
Slow-moving invertebrates, including small slugs, earthworms and snails

BREEDING
Breeding season: April–August; gestation period: 90–100 days; number of young: 4 to 22; breeding interval: usually 2 years

LIFE SPAN
Up to 55 years

HABITAT
Large range of habitats, but not land used for grazing or farming, or dense woodland

DISTRIBUTION
Scotland and northern Spain in the west to Ural and Caucasus Mountains in the east

STATUS
Common

Slowworm

When in the open slowworms risk capture by a long list of predators, including hedgehogs, adders, frogs and toads, birds of prey, foxes and badgers.

SLUG

The banana slug is found on moist forest floors on the Pacific coast of North America from California to Alaska. It eats a range of foods, particularly mushrooms. In doing so, it plays an role in distributing fungal spores.

DURING THE EVOLUTIONARY history of land snails, members of certain groups lost their external shells. Although these animals are not necessarily all closely related, we call them all slugs. In fact, some species of slug retain a vestigial (imperfectly developed) shell, usually hidden within the body. Slugs belong to the Pulmonata, a large group of land and fresh-water gastropods that breathe with a lung.

There are numerous types, classified into many families. The keelback slugs, roundback slugs and shelled slugs, for example, form the families Limacidae, Arionidae and Testacellidae, respectively. The first are named after the ridge or keel on the upper body toward the hind end. Behind the head with its four tentacles, the back is covered by a roughly elliptical shield formed from the mantle, the part of the body wall that secretes the shell material in shelled mollusks. The mantle is pierced by the breathing pore on its right margin. The tiny shell of the keelback slugs is a flattened oval, horny and poorly calci-fied, hidden under the mantle shield. In the roundback slugs the shell is usually reduced to a number of separate chalky granules. The North American roundback slug *Binneya* sp. is an exception, because it has an external spiral shell. The largest of the roundback slugs may be 8 inches (20 cm) long, as in the banana slug, *Ariolimax columbianus*, of North America.

In the third group, the shelled slugs, there is a small shell visible on the surface toward the rear of the animal. The mantle, heart and single kidney of a shelled slug lie under the shell toward the broad rear end. A groove runs forward from the mantle on either side of the body, giving off branches to the back and flanks.

Silver trails

Relative to snails, slugs are at greater risk from drying out and predation. However, the animal's load is light compared to that of a snail, and its need for calcium to form a large, mineralized shell is much less. It can also creep through much smaller holes than a snail can, and so can move effectively underground. Slugs are confined to damp, secluded places; in fact, some species spend most of their lives under-ground. They are most active at night, emerging by day only when it is wet. Although most feed near ground level, some are good climbers and regularly ascend trees to heights of 30 feet (9 m) or more. The tree slug, *Lehmannia marginata*, and gray field slug, *Deroceras reticulatum*, are two such climbers, their silver trails of slime running up and down some tree trunks attesting to these activities. These slugs may take a more rapid route for the descent and lower themselves many feet through the air on a string of slime. Such slugs spend the day in knotholes or other cool, secluded places, coming down to the ground at dusk and climbing up again at dawn.

Not only lettuce

Hated by gardeners, slugs may be much more numerous in their gardens than people realize. Little of slugs' food consists of the plants gardeners cultivate, except where there are few alternatives. Instead, some species of slugs feed almost entirely on fungi, eating little or no vege-tation and then only when it is dying or rotting. Many slugs are omnivorous and are attracted by fungi, vegetation, tubers, carrion, dung, kitchen refuse or the metal-dehyde-baited bran put out to kill them. They are drawn to such foods over distances of several feet by the odor, detected by the slug's organs of smell situated in its tentacles. In confinement, slugs in general may turn on each other, but the shelled slugs are particularly notable for their predatory habits. They are most common in well-manured gardens and live underground most of the time. They feed by night on earthworms and to a lesser extent on centipedes and other slugs. A slug feeds using its *radula*, a specialized feeding organ a little like a conveyor belt covered with lots of tiny teeth. The slug draws the radula back and forth over its food stuff, abrading it and loosening morsels of it, which it draws towards its gullet.

SLUGS

PHYLUM **Mollusca**

CLASS **Gastropoda**

SUBCLASS **Pulmonata**

ORDER **Stylommatophora**

FAMILY **Numerous, including Limacidae, and Arionidae and Testacellidae**

GENUS AND SPECIES **Thousands of species and numerous genera including, Limacidae: *Limax, Ariolimax, Lehmannia, Deroceras*; Arionidae: *Binneya, Arion*; and Testacellidae: *Testacella***

LENGTH
Less than ½ in. (1 cm) to 8 in. (20 cm)

DISTINCTIVE FEATURES
Coated in slime; various sizes and colors; 4 antennae; long, cylindrical bodies; strong, muscular foot; slightly raised region toward front of body, called a mantle; feeds with radula

DIET
Many species: plants only; some species: fungi and soil; various decaying plant, animal or fungal material

BREEDING
Generally hermaphroditic, but young slugs are male. Fertilization usually mutual, but self-fertilization possible.

LIFE SPAN
Variable; up to several years

HABITAT
Moist and humid environments; some species common in gardens and cultivated crops; some species entirely underground

DISTRIBUTION
Worldwide, but confined to moist environments

STATUS
Abundant to not known

Aerial courtship

Slugs are hermaphrodites, having both male and female parts in the same individual. Although self-fertilization can occur, a two-way exchange of sperm between mating pairs is usual. In the first stages of mating, roundback and keelback slugs typically circle, leaving a trail of chemical stimulants and attractants in their slime They constantly contact each other and devour each other's slime until they come to lie side by side. The great gray slug, *Limax maximus*, concludes this circling in a particularly spectacular manner. Climbing first up a tree or wall, the two slugs circle for a period of ½–2½ hours, flapping their mantles and eating each other's slime. Then suddenly they wind spirally around each other before launching themselves heads downward into the air on a thick cord of slime, perhaps 18 inches (46 cm) long. Now the penis of each is unrolled to a length of 2 inches (5 cm) and entwined with the other into a whorled knot. Sperm masses are exchanged, after which the slugs either fall to the ground or reascend their lifeline, eating it as they go. The eggs are laid soon afterward in a damp recess such as under a stone or among roots. The soft amber eggs, about ⅕ inch (5 mm) across, hatch in about a month.

Despite their unpleasant slime, slugs are eaten by a variety of predators, including frogs, toads, hedgehogs, ducks, blackbirds, thrushes and other birds. Ducks are especially good at controlling the numbers of slugs. Slowworms and various insects also take their toll. Although sheep are not deliberate predators of slugs, they do eat them accidentally and in doing so may become infected with lungworm, a parasitic nematode, whose larvae have formed cysts in the foot of the slug.

Because slugs are hermaphrodites, a mating between two slugs results in both members of the couple laying fertilized eggs.

SMELT

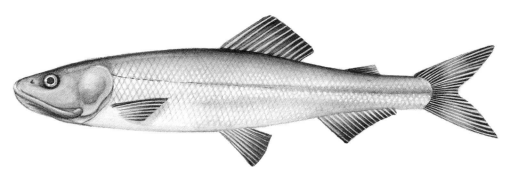

The European smelt is found in the eastern Atlantic Ocean and connected bodies of water between 38° N and 70° N. It may live for 10 years and grow to a maximum length of 18 inches (45 cm).

SMELTS ARE SMALL, SILVERY fish that seem to live equally well in the sea and in estuaries. They are remarkable for their large numbers and many are taken by larger fish.

The European smelt, *Osmerus eperlanus*, is long and slim, with a pointed head and silvery body, olive green on the back with a slight bluish green tinge to the fins. The jaws are large, the lower jaw jutting beyond the upper. There are fine teeth in the jaws, larger in the lower jaw than in the upper jaw, conical teeth on the roof of the mouth and several large fanglike teeth on the tongue. The dorsal fin is set far back, and there is a small adipose (fatty) fin just forward of the slightly forked tail fin.

The European smelt is found from the River Seine to the Baltic Sea. A similar species lives off the Atlantic coast of North America, from the Gulf of St. Lawrence south to Virginia. The eulachon, *Thaleichthys pacificus*, another species of smelt, which grows to 12 inches (30 cm), lives on the Pacific coast from Alaska to Canada, and the surf smelt, *Hypomesus pretiosus*, ranges from Alaska to California. One of the smallest smelts is the 3-inch (7.5-cm) longfin smelt, *Spirinchus thaleichthys*, of the San Francisco area. The remaining species of smelt live in the North Pacific. There are no smelts in the Southern Hemisphere.

Freshwater and saltwater fish

Smelts live in large shoals in coastal waters and estuaries and are rarely found far from the shore. In European waters some smelts spend their whole life in the larger estuaries. Young smelts are often found in pools between tidemarks.

In parts of Europe, smelts have become permanently landlocked in fresh water. This has also occurred in North America. The Atlantic smelt was introduced into parts of the Great Lakes and became numerous enough to be fished commercially. In California there is a freshwater smelt that spends almost its whole life in the Sacramento River. However, the same species in Japan lives in the sea and only migrates into rivers to spawn.

Eggs with sticky flaps

The shoals of adult European smelts congregate around large estuaries in late winter to enter the rivers for spawning. This begins in March in the waters around Britain, later on parts of the Continent. Smelts are caught commercially during these spawning runs; the males develop small tubercles on their scales at these times. After spawning they return to the sea, but the young, hatching from the eggs, remain in the estuaries until the end of the summer. Development follows the normal pattern for this type of fish, but the eggs are distinctive. They are pale yellow, 1.3 millimeters in diameter and sink to the bottom. The eggs are enclosed in a double membrane, and as they sink the outer layer breaks away in part and the loose part is turned back. Its inner surface is sticky and adheres to stones and other hard objects, anchoring the eggs, which hatch 8–27 days later, depending on the water temperature. Each newly hatched larva is ¼ inch (6.4 mm) long but grows to nearly 3 inches (7.5 cm) by the end of the first year.

Smelts feature in the food chains of larger fish-eating species, and humans have fished smelts for centuries. Early Native Americans caught the eulachon in large numbers. They ate the fish itself, but also made use of its flesh, which is especially oily in the breeding season. The fish was dried and, when tied to a stick, could be lighted and used as a torch.

Predatory fish

Smelts have numerous teeth and are predatory fish. The young smelts feed on tiny crustaceans, especially copepods, and on fish larvae. They also take small worms. They soon graduate to taking young herring and the young of other fish, such as sprats, whiting and gobies, as well as a variety of crustaceans, from copepods to shrimps.

SMELTS

CLASS	**Osteichthyes**
ORDER	**Osmeriformes**
FAMILY	**Osmeridae**
GENUS	**6 genera**
SPECIES	**15 species, including eulachon,** *Thaleichthys pacificus* **(detailed below)**

LENGTH
Up to 12 in. (30 cm)

DISTINCTIVE FEATURES
Long, slender body; large jaws, extending back past eyes; lower jaw longer than upper jaw; teeth on both jaws; 2 dorsal fins

DIET
Plankton; only feeds while at sea

BREEDING
Spawning run from sea to freshwater streams begins when river temperature rises to about 39.9° F (4.4° C); fish stop running if temperature exceeds 46° F (7.8° C). Adults usually die after spawning; some move back to sea and return to spawn again. Hatching period: 8–27 days. Upon hatching, larvae move to near bottom; soon carried downstream to salt water.

LIFE SPAN
Up to 5 years

HABITAT
Near shore; coastal inlets and rivers, perhaps to depths of 2,060 ft. (625 m); returns to spawn in freshwater streams

DISTRIBUTION
North Pacific: west of Saint Matthew Island and Kuskokwim Bay in Bering Sea and Bowers Bank in Aleutian Islands south to Monterey Bay, California.

STATUS
Not threatened

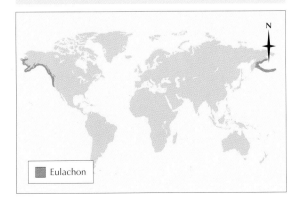

Eulachon

Disappearing trick with mirrors

There is an apparent contradiction in a fish being widely preyed upon yet continuing to exist in teeming numbers. Animals living in large herds on land or birds flying in large flocks enjoy safety in numbers. A predator attacking them tends to be confused by their numbers, and has to single out its victim and cut it off from the rest in order to make a kill. However, ichthyologists reason that a solitary fish needs less efficient camouflage than fish living in shoals because it presents such a small target compared with a shoal.

Fish such as dace live in clear water and their silvery sides reflect their surroundings like a mirror, so that from any angle they merge with their background and are difficult to see. By contrast, smelts live in slightly dirty waters close inshore or in estuaries, where the light is scattered. Under their scales are platelets of crystals and the angle these make with the surface ensure that a smelt swimming in the normal position can only be seen in silhouette by anything regarding it from immediately below. From any other angle the light reaching the eye from the platelets has the same intensity as the light coming from behind the body, effectively rendering the fish nearly invisible.

These smelts were caught off the coast of Washington State. Tons of smelts are caught each year on their spawning runs in the Columbia River.

SNAKEHEAD

Snakeheads, such as this Channa lucius, are said to make good pets. However, they need enormous tanks or ponds, which must be heated to tropical temperatures.

SNAKEHEADS ARE RATHER distinctive freshwater fish of Africa and southern Asia. Although included in the huge and diverse order Perciformes, their relationships to other fish within this group are obscure. They have been linked with the gouramis, which might be their closest relatives. The two genera of snakeheads, *Channa* and *Parachanna*, are therefore put in their own family, the Ophicephalidae.

Snakeheads have long bodies, cylindrical in front, but slightly compressed from side to side toward the hind end. They have large reptilian heads with a jutting lower jaw and numerous teeth. The head is so massive and the jaws are so long that its gape can be enormous. There is a pair of double nostrils. The dorsal fin is soft-rayed and runs from just behind the head almost to the tail. There is a similarly long anal fin. The body is usually mottled with brown, sometimes with tinges of red. There are often distinctive gray, brown and black markings along the body, V-shaped in the African species. These markings amount to a disruptive camouflage pattern that makes the fish blend with their background as they rest among the water plants. The smallest species, *Channa gachua* is about 6⅔ inches (17 cm) long, whereas most species are up to 3 feet

(90 cm) long. In the larger species, for example the giant or Indian snakehead, *C. marulius*, exceptional individuals grow to 6 feet (1.8 m). The Asian species range from Sri Lanka through Pakistan and India to northern China and Southeast Asia.

Fishing with knives

Snakeheads can live in foul and stagnant waters that contain too little oxygen to support most other fish. When it cannot get the oxygen it needs from the water, a snakehead comes to the surface to gulp in air. Inside the gill chamber on each side are tiny pouches that amount to accessory air-breathing respiratory organs. They are well supplied with a network of fine blood vessels that take up oxygen. When the water in which it lives becomes even more stagnant, or begins to dry up, a snakehead can escape completely by traveling overland to a more favorable body of water. To move on land, it wriggles its body and makes rowing movements with its short, broad pectoral fins. Another way in which snakeheads cope with their ponds drying up is to bury themselves in the mud to a depth of 1–2 feet (30–60 cm). It is then that human hunters, armed with knives, slice at the mud to find them.

GIANT SNAKEHEAD

CLASS	**Osteichthyes**
ORDER	**Channiformes**
FAMILY	**Ophicephalidae**
GENUS AND SPECIES	***Channa marulius***

ALTERNATIVE NAMES
Indian snakehead; great snakehead

WEIGHT
Up to 66 lb. (30 kg)

LENGTH
Up to 6 ft. (1.8 m)

DISTINCTIVE FEATURES
Large head; mouth very deeply cleft, with huge gape; complete set of teeth; very long-based dorsal and anal fins, without spines; accessory respiratory organ capable of breathing air; blotched pattern on skin, but predominantly ochre, fawn or pale gray-brown in color

DIET
Fish, frogs, snakes, insects, earthworms and tadpoles

BREEDING
Age at first breeding: 1–2 years; breeding season: April–June in Sri Lanka; number of eggs: 500; hatching period: 30–54 hours

LIFE SPAN
At least 8 years

HABITAT
Lakes and deep pools in rivers, including those with very poor water quality and little oxygen

DISTRIBUTION
Southern Asia, from Pakistan to northeastern China in north, Vietnam in east, and Malaysia and Sri Lanka in south

STATUS
Locally common

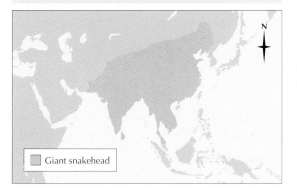

Giant snakehead

Lurking in the weeds

When first hatched, snakeheads feed on small plankton such as water fleas, rotifers and crustacean larvae. As they grow, they take larger plankton including insect larvae, and when about 2 inches (5 cm) long they begin to feed on small fish, as well as insects and their larvae. When mature, they feed mainly on other fish, which they stalk stealthily, approaching them from the front. Then they bend themselves into an S-shape and throw their heads forward with a sudden jerk to seize their prey. Adult snakeheads are formidable creatures; they eat frogs and even tackle water snakes.

Sheltered larvae

Snakeheads begin to breed when about a year old. A pair of snakeheads choose a shallow, sheltered part of a slow moving river or lake, and clear the water plants over a small area. The females then shed their eggs into the water almost at random and the males do the same with their milt. The eggs contain oil droplets so they rise to the surface and float. The giant or Indian snakehead parents together construct a cup-shaped floating nest for the eggs from weeds and leaves. The eggs hatch in 2–3 days, the larvae continuing to feed on the yolk sac for another 6–8 days, during which time they float belly-up on the surface. After this they are able to swim normally and they grow fairly rapidly, but they make frequent visits to the surface to gulp air, and where they are numerous, the water appears colored as a constant procession of larvae rise to the surface. In some species the males guard the eggs and the larvae for a while, but in the giant snakehead, both parents zealously guard the young fish.

Once the juveniles are able to swim on their own, they hide among the water plants, a habit that then continues throughout their lifetime because snakeheads, like the pike, spend much of their time concealed among vegetation, from which they emerge to seize their prey.

Fresh fish

Snakeheads are important food fish in southern Asia and are bred in irrigation wells in India. They have also been introduced into parts of the United States, where they are now flourishing. Because they are air-breathers, they can stay alive for days on end out of water and so remain fresh until they are sold or needed for cooking. This is important in areas where few families have a refrigerator. Snakeheads are said to be able to live for some time out of water, breathing air and deriving nourishment from the fat stored in their bodies. In the fish markets they are put out for sale live, on woven trays.

SNAKE MACKEREL

The snake mackerel, Gempylus serpens, needs long, fanglike teeth to restrain energetic prey such as flying fish.

SNAKE MACKEREL ARE OCEANIC fish living in moderately deep waters and are distantly related to true mackerel. The name snake mackerel is sometimes given to any member of the snake mackerel family, a family that includes the gemfish, oilfish and escolars, or it may apply to particular species, for example, *Gempylus serpens*. The snoek, also described in this volume, is another member of the snake mackerel family. Two snake mackerel leaped to fame in 1947 when they jumped onto the famous raft of Thor Heyerdahl, the *Kon-Tiki*. This moment of lime-light was unusual for an interesting group of fish that is rarely placed under enough scrutiny.

Snake mackerel have long, fairly slender bodies and a long snout with a prominent lower jaw that ends in a pointed fleshy tip. The mouth is armed with fanglike teeth. The eyes are large. The body is covered with small, smooth scales and small, forked spines. The first dorsal fin, which begins just behind the head, is long and becomes gradually lower toward its hind end. It consists of spines joined together by membranes. Just behind it is a second dorsal spine followed by a few finlets, each of which consists of an isolated spine with a membrane. The anal fin is also followed by a few finlets. The pectoral fins are small, and the pelvics are very tiny. The back is dark brown to black and there are violet tints on the flanks and belly. There are two lateral lines, one just below the dorsal fin and the other along the midline of the flank. Both are whitish in contrast to the fins and inside of the mouth, which are black.

One species of snake makerel, *Nesiarchus nasutus*, also known as the black gemfish, grows to a length of 4 feet (1.3 m) and is found in the North Atlantic from near the surface down to 3,600 feet (1,100 m). *Gempylus serpens*, which grows to 5 feet (1.5 m), is found in all tropical seas and dives to a similar depth.

Daily vertical journeys
Like so many deep-sea fish, snake mackerel make daily vertical migrations, the adults appearing near the surface at night. There is no special effort made by the commercial fishing industry or by sport anglers to catch snake mackerel. We learn what we can about them when they are caught as by-catch when fishing for tuna by longline. Examination of the stomach contents of such fish reveals that *Nesiarchus nasutus* feeds on the viperfish of genus *Chauliodus*, snipe-eels and squid, while *Gempylus serpens* seems to feed on flying fish. Individuals have been witnessed catching flying fish, and the snake mackerel landing on the deck of the *Kon-Tiki* were probably engaged in hunting this type of prey.

SNAKE MACKEREL

CLASS	**Osteichthyes**
ORDER	**Perciformes**
FAMILY	**Gempylidae**
GENUS	**16 genera**
SPECIES	**23 species, including *Gempylus serpens* (detailed below)**

LENGTH
Up to 3⅓ ft. (1 m)

DISTINCTIVE FEATURES
Large, very elongated fish; long jaws; strong front canines; body covered with small, smooth scales and small, forked spines; dorsal and anal fins followed by finlets; dark brown to black back; violet tints on flanks and belly

DIET
Fish, squid and crustaceans

BREEDING
Males mature at 17 in. (43 cm); females mature at 20 in. (50 cm). Breeding season: year-round, in tropical waters; number of eggs: about 300,000 to 1 million; eggs are planktonic

LIFE SPAN
Not known

HABITAT
Warm oceans; adults migrate to surface at night; larvae and juveniles stay near surface only during day

DISTRIBUTION
Worldwide in tropical and subtropical seas; adults also found in temperate waters

STATUS
Not known

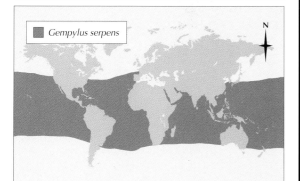

Gempylus serpens

Year-round spawning

Gempylus serpens is believed to spawn throughout the year, and two spawning areas have so far been located, one in the Caribbean and one off Florida. The larvae and juvenile fish also undergo daily vertical migrations in the water column, but in the opposite direction of the adults' migrations. The young fish appear near the surface during the day, then dive down to the depths at night.

Medicine fish

Another member of the same family is the escolar, *Ruvettus pretiosus*, also known as the oilfish, scourfish, or castor-oil fish. The name escolar may come from the Spanish word meaning to scour or burnish, which may refer to the roughness of the skin. It is deeper bodied than the others and has a large but less specialized mouth. It is found in all tropical oceans and the Mediterranean, at depths of 600–2,400 feet (180–730 m). The fishers of Madeira and the Canary Islands that commonly use longline fishing catch it. Its flesh is very oily. It is called the castor-oil fish because its flesh acts as a purgative.

While the snake mackerel are considered by the food market as fit only for sausages and fish cakes, the closely related cutlassfish are valued as commercial species in most parts of the world. They are even used for sashimi in Japan. Nevertheless, in the United States, anglers use the cutlassfish they catch as bait for larger, more prized sport fish. Cutlassfish are elongated like snake mackerel, but their bodies taper to a pointed tail without a tailfin. They are not strictly pelagic like the snake mackerel, but prefer shallow water and muddy bottoms, especially during the day, when they are known to gather in large numbers in bays and estuaries. By night they hunt near the surface, and catch sardines, anchovies, squid and crustaceans.

Cutlassfish such as this Trichiurus nitens can reach 5 feet (1.5 m) long and 7 pounds (3.2 kg) in weight. This preserved specimen did not reach such a size.

SNAKE-NECKED TURTLE

SEVERAL AUSTRALIAN SPECIES of turtles have outlandishly long necks. The combined length of the head and neck can be greater than that of the shell. These turtles are members of the suborder Pleurodira, all members of which bend their neck sideways when they withdraw their head into the shell. Pleurodirans are often called side-necked turtles, hidden-necked turtles or snake-necked turtles. These three names tend to be used interchangeably (see hidden-necked turtle). All other turtles, including the terrestrial tortoises, belong to the Cryptodira, and withdraw their head in a vertical plane. Unlike some cryptodires, the pleurodires cannot withdraw their heads out of sight. It is the eight species of extremely long-necked Australian pleurodires that are most legitimately called snake-necked turtles, and will be the subject of this article.

The most well-known species is often called the common snake-necked turtle, but can also be referred to as the eastern snake-necked turtle or long-necked tortoise. It is found from central Queensland to the south coast of Victoria in Australia. These turtles have a rather flat, brown or black shell, but in some pale brown specimens each of the scutes (bony plates making up the shell) is outlined in dark brown or black. Hatchlings sometimes have red or orange-colored shells. They are comparatively small; the biggest specimens have a shell that is 11 inches (28 cm) in length.

Built-in snorkels

Snake-necked turtles live in fresh water, in rivers, ox-bow lakes, ponds and swamps. Most species have a long, pointed snout and this enables them to breathe with only a very small amount of the head showing above the surface of the water. They are not strong swimmers, and prefer to walk along the bottom. In this context, their long necks act as snorkels, enable them to reach the air at the water surface from their resting place on the river bed. Many of the Australian species spend most of their time in the water, but common snake-necked turtles sometimes make extensive migrations over land, especially if their normal watery habitat is in danger of drying up.

The common snake-necked turtle, Chelodina longicollis, *is widespread in Australia. It leads an aquatic life but has the ability to migrate across land when its water habitat begins to dry up.*

SNAKE-NECKED TURTLES

CLASS	**Reptilia**
ORDER	**Testudines**
SUBORDER	**Pleurodira**
FAMILY	**Chelidae**

GENUS AND SPECIES   **8 species including common snake-necked turtle, *Chelodina longicollis*; and broad-shelled river turtle, *C. expansa***

ALTERNATIVE NAMES
C. expansa: Murray River turtle; giant snake-necked turtle. C. longicollis: eastern snake-necked turtle; long-necked tortoise.

LENGTH
**C. expansa: up to 20 in. (50 cm);
C. longicollis: up to 11 in. (28 cm)**

DISTINCTIVE FEATURES
Flat-topped shells, very long necks withdrawn by bending neck sideways

DIET
Exclusively animals: mollusks, crustaceans, insects, tadpoles and small fish

BREEDING
Breeding season: eggs laid March–May; number of eggs: 5 to 25; hatching period: 3–12 months

LIFE SPAN
Not known

HABITAT
Wide variety of freshwater habitats: slow-moving rivers, ponds, ox-bow lakes, streams and swamps. Temporary streams, dry for long periods each year (some species only).

DISTRIBUTION
Most of Australia except Tasmania and the central deserts; New Guinea

STATUS
Varies from fairly abundant (most species) to not known (some species)

Snake-necked turtles (genus *Chelodina*)

Pile-driving female

Some of the snake-necked turtles have restricted distributions in remote parts of tropical Australia and inaccessible New Guinea. Details of breeding and other biology are known only in common Australian examples such as the broad-shelled river turtles that live in the Murray and Darling Rivers and their tributaries. Courtship involves head bobbing by individuals of both sexes. In most parts of their range, the females lay their eggs from March to May, which is autumn in Australia. Each female crawls out of the water and may travel for up to 200 yards (180 m) in search of suitable sandy soil in which to lay her eggs. Once satisfied, she digs a conical hole about 8 inches (20 cm) deep in which she lays her eggs, usually 5 to 25 in number. The female uses her hind legs to fill the hole with soil, and then tamps it down with her body, which is raised and dropped like a pile driver. Once she has finished, the nest is almost impossible to find.

The incubation period in broad-shelled river turtles is very variable. Hatching may occur in as little as 3 months, or it may not take place until more than a year after the eggs have been laid. The baby turtles sometimes remain in the nest for several weeks after they have hatched from the eggs. Finally, to add to the variability, some females lay their eggs in spring, not in the fall.

The snake-necked turtles are related to other Australian pleurodiran, or side-necked turtles, such as this Murray turtle, Emydura maquarii.

SNAKES

*S*NAKES LIVE THROUGHOUT the world's tropical, subtropical and temperate regions, with some able to survive high up in the Himalayas or as far north as the Arctic Circle. Most snakes are terrestrial, but others have colonized rivers and seas, while still more are adapted to life in the forest canopy. The group evolved over 100 million years ago, yet some snakes retain the remnants of their pelvic girdle, which once held the hind limbs of their ancestors.

Classification conundrums

Snakes belong to the order Squamata, along with the lizards and worm lizards, or amphisbaenas. They are placed in a separate suborder: the Serpentes, to distinguish them from the lizards and their kin, placed in the suborder Sauria. Members of the Serpentes lack limbs and move around by undulations of their long, thin bodies, producing characteristic S-shaped curves. Snakes can be distinguished from other legless members of the Squamata, like legless lizards, by their lack of eyelids and the overlapping pattern of their scales.

The Serpentes are divided into the Scolecophidia and the Alethinophidia. Members of the Scolecophidia are small, primitive, burrowing creatures, with vestigial (imperfectly developed) eyes that are often covered with scales. Many have

The emerald tree boa, Corallus caninus, *of the Amazon forest, drapes itself over horizontal branches and lies in wait for prey.*

tiny claws at the base of their tails where their ancestors once had legs. The infraorder contains three families, members of which lack venom. The Alethinophidia contains all the other snakes and can be divided into two superfamilies, the Henophidia and the Xenophidia. The former contains the constrictors, including the pythons and boas, while the latter contains all the well-known families of venomous snakes, including the vipers (Viperidae) and cobras (Elapidae).

The taxonomy of snakes is still the source of much controversy. For instance, the

CLASSIFICATION	
CLASS	Reptila
ORDER	Squamata
SUBORDER	Serpentes
INFRAORDER	Scolecophidia: primitive snakes; Alethinophidia: other snakes
NUMBER OF SPECIES	Approximately 2,700

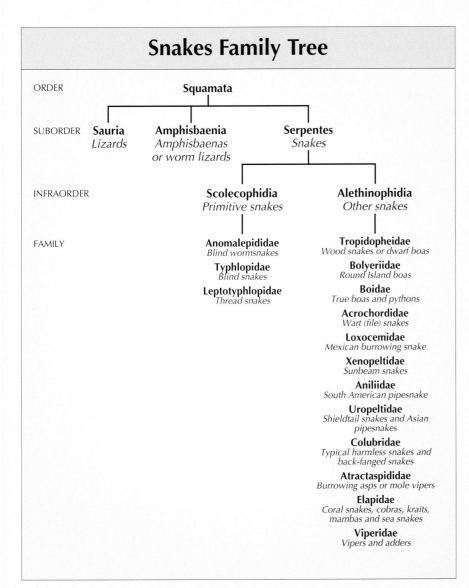

Snakes Family Tree

ORDER		**Squamata**	
SUBORDER	**Sauria** *Lizards*	**Amphisbaenia** *Amphisbaenas* *or worm lizards*	**Serpentes** *Snakes*
INFRAORDER		**Scolecophidia** *Primitive snakes*	**Alethinophidia** *Other snakes*

Scolecophidia *Primitive snakes*

Anomalepididae
Blind wormsnakes
Typhlopidae
Blind snakes
Leptotyphlopidae
Thread snakes

Alethinophidia *Other snakes*

Tropidopheidae
Wood snakes or dwarf boas
Bolyeriidae
Round Island boas
Boidae
True boas and pythons
Acrochordidae
Wart (file) snakes
Loxocemidae
Mexican burrowing snake
Xenopeltidae
Sunbeam snakes
Aniliidae
South American pipesnake
Uropeltidae
Shieldtail snakes and Asian pipesnakes
Colubridae
Typical harmless snakes and back-fanged snakes
Atractaspididae
Burrowing asps or mole vipers
Elapidae
Coral snakes, cobras, kraits, mambas and sea snakes
Viperidae
Vipers and adders

family Boidae is considered by some taxonomists to contain both the boas and the pythons, while others place the pythons in a separate family, the Pythonidae. The boas and pythons include the largest snakes: both the reticulated python, *Python reticulates*, and the anaconda, *Eunectes murinus*, can reach lengths of 30 feet (10 m). The Boidae also contains some more moderate sized snakes, although the dwarf boas, which were once its smallest members, have recently been placed in a separate family.

Other snakes have also proved difficult to classify. The Mexican burrowing snake, *Loxocemus bicolor*, was once thought to be the only New World boa but is now the only representative of a new family, the Loxocemidae, whose relationships to other families is uncertain. The file snakes (family Acrochordidae) are another such enigma. Members of the group are totally aquatic, living in coastal waters and estuaries, and have a scale pattern completely unlike that of even their closest relatives.

Perhaps the largest-scale taxonomic problem is the family Colubridae, by far the most extensive snake family, with more than 1,500 species. Most taxonomists expect the family to be divided in future, as scientists learn more about them. Members of the

The horned viper, Cerastes cerastes, ***is one of the sidewinding vipers. Vipers can exploit apparently hostile environments that other snakes cannot.***

Colubridae are generally long and slender, with large scales and big eyes. Although some are venomous, few species are dangerous to humans.

Most of the world's most dangerous snakes are members of the Elapidae and the Viperidae. The elapids are recognized by their hollow fangs, which are fixed to the front of their upper jaw and connected to specialized venom ducts. The group contains the world's largest venomous snake, the king cobra, *Ophiophagus hannah*, which can reach lengths of up to 15 feet (5 m). The vipers are generally considered the most evolutionarily advanced snake family. They seem to have evolved after the Australian landmass had broken away, and so are not represented on that continent. Otherwise, however, they are spread worldwide, and can often inhabit areas too inhospitable for other snake groups.

Sixth and seventh senses

Snakes employ an unusual array of senses to detect their prey. They rely primarily on their acute sense of smell, in combination with a related sense associated with the Jacobson's organ, found only in snakes and a few lizards. The organ consists of two tiny sacs lined with sensory cells, which open onto the roof of the mouth via narrow ducts. Snakes draw airborne molecules from their environment to the organ by flicking their tongue and then process the information in the olfactory lobe of the brain, the same part that processes smell signals.

A grass snake, **Natrix natrix,** *tastes the air, then flicks its tongue over its Jacobson's organ to process the information.*

Pit vipers and some species of pythons and boas possess a sense that is unique in the animal kingdom. They have specialized heat-sensitive pits in the side or front of the head, which they use for seeking out warm-blooded prey. Pit vipers possess the most sophisticated heat sensors, with some species able to detect temperature changes of as little as 0.002° F (0.001° C), and to accurately locate the heat source. Using their specialized organs, they effectively see their environment in terms of radiated heat (infrared waves and microwaves). The heat sense of snakes far surpasses that of other heat-sensing hunters, such as the vampire bat.

In comparison, snakes' vision is not well developed, which for a vertebrate predator is unusual. With the exception of one genus, *Ahaetulla*, snakes cannot focus by changing the shape of their lenses, only by moving them backwards and forwards like a camera. This method is slow, and provides limited focusing ability.

The auditory (hearing) system of snakes is also limited, although they are not, as is popularly believed, deaf. Although they lack visible ears, snakes possess the remnants of sound-processing equipment in the form of a small bone called the stapes, which is indirectly connected to the lower jaw. Since their jaws will often be in contact with the ground, snakes are likely to be extremely sensitive to ground vibrations.

Eggs and live young

Snakes are extremely varied in their reproductive strategies. In temperate species, both sexes have an annual breeding cycle. Males only produce sperm during summer and fall, when

food is plentiful, while a female's eggs usually ripen in spring, just before breeding season starts. Breeding cycles are less obvious in tropical species. Some may react to subtle changes in temperature associated with the onset of the rainy or dry seasons, while other may breed all year round.

After fertilization the majority of snakes lay eggs. The clutch sizes of different species vary enormously, from fewer than 10 eggs in coral snakes to more than 100 in large pythons. Snake eggs need to be kept warm and moist, so the female often chooses her nest site carefully. Female pythons incubate their eggs, coiling their bodies around them and generating heat through muscular contractions.

Although egg laying is the most common strategy, many snakes are viviparous, bearing live young. Viviparity has advantages in some habitats. Sea snakes (subfamily Hydrophiinae), for example, clearly benefit from giving birth to swimming young, rather than having to return ashore to breed. Many arboreal (tree-living) species are also viviparous. When born, their young are immediately capable of moving about the canopy.

The primitive thread snake *Ramphotyphlops braminus*, or flowerpot snake, has a bizarre method of reproduction. All flowerpot snakes are female and can breed without the help of a male, since their eggs develop without the need for fertilization. This female-only reproduction is known as parthenogenesis; known in leeches, flatworms, aphids and other invertebrates, it is extremely rare among vertebrates.

Predators

All snakes are predatory. The Alethinophidia typically feed on large prey such as vertebrates, eggs and large invertebrates. However, the Scolecophidia, the small, primitive, burrowing worm snakes and blind snakes, eat small prey such as ants, termites, earthworms and other small, invertebrate inhabitants of the soil.

Many snakes are active diurnal (daytime) hunters, relying on speed and agility to hunt down prey during the warmth of the day, when they are able to be most active. Colubrids such as the whip snakes (genus *Coluber*) are examples of such hunters. Some snakes are foragers, stealing eggs from nests or feeding on slow-moving invertebrates, while others are sit-and-wait predators. One such snake, the Brazilian viper, *Bothrops bilineata*, uses its colorful tail to lure in its prey.

Once located, many snakes rely on venom to overpower prey. Colubrids have a primitive venom-delivery system, in which a modified salivary gland supplies venom to fangs at the back of the mouth. By contrast, the front-fanged elapids and vipers are able to inject their venom through narrow grooves, transferring it much more efficiently to their prey.

Constrictors lack venom. Instead, they restrain prey in their coils, causing suffocation or preventing the heart from beating. All snakes swallow their prey whole. Large prey items

To hatch, a baby snake such as this eastern hognose snake, Meterodon platyrhinos, *must slit the parchmentlike shell of its egg with its egg tooth.*

are swallowed headfirst, since the limbs of vertebrates fold more easily in this direction. The snake opens its mouth wide and hooks the teeth on one side of its jaw into the prey. It then slides the other side of its jaw forward, hooks it into place and repeats the process. They can dislocate the bony joints between the upper and lower jaws and between the two separate halves of the lower jaw, so that their jaws are held together only by ligaments. In doing so, snakes are able to stretch their jaws apart to an remarkable degree, to accommodate prey many times the diameter of the head.

Persecution

Snakes have not escaped the general decline in wild predator populations. Habitat destruction is likely to be a major cause of the fall in snake numbers, and snakes have suffered badly from the effects of deforestation. Road deaths also take a heavy toll. The slow-moving and elongated bodies of snakes are easy targets for the careless or malicious driver.

The rhinoceros viper, Bitis nasicornis, *is an ambush predator.* It remains unnoticed until prey come within striking distance.

Snakes are heavily exploited for their skins, which support a lucrative trade in exotic handbags, shoes and wallets. Most snakes killed in this way originate in East Asia and are sold to buyers in North America, Europe and Japan. Several large constrictors, including two species of pythons, *Python reticulates* and *P. molurus*, and the common boa constrictor, *Boa constrictor*, are among those species most heavily exploited, and between them support an industry worth several million dollars a year in the United States. The attractive Indian rat snake, *Ptyas mucosus*, is another species to have been hit hard by the snakeskin industry and has shown a marked decline in numbers. It seems certain that the present levels of exploitation are unsustainable, and may lead to a crash in snake populations.

There is also a thriving trade in live animals, both for the pet trade and for display in zoos and private collections. Fortunately, many of the more popular species are now bred in large numbers by commercial breeders. More exotic species of snake cannot be obtained from breeders, and captured animals continue to account for a significant number of the snakes removed from the wild.

In addition to commercial exploitation, snakes may also be killed simply because they are considered a threat. In the United States, early settlers assiduously hunted down as many rattlesnakes as possible, in the interests of safeguarding themselves and their livestock. Rattlesnake round-ups became popular sport in some areas, and such spectacles continue to this day in a few states. In Sweetwater, Texas, 70,773 snakes were reportedly killed over 16 years. Culls of this size must threaten snake populations.

Many species of snakes are protected by law, but in most cases legislation is not sufficiently specific or far-reaching. Laws exist to prevent the capture or deliberate extermination of a number of species in Europe and the United States, but in most cases they cannot prevent habitat destruction or road deaths. A few subspecies are afforded more stringent protection, including the San Francisco garter snake, *Thamnophis sirtalis tetrataenia*, and the New Mexico ridge-nosed rattlesnake, *Crotalus willardi obscurus*, in the United States, and *Macrovipera schweizeri*, a viper found only in Greece.

The I.U.C.N. (World Conservation Union) currently lists five species of snakes and one further subspecies as endangered, and considers another five species to be vulnerable. But many more species are not classified because of a lack of accurate population data, and the true scale of the problem is likely to be much greater.

For particular species see:
- ADDER • ANACONDA • ASP • BOA CONSTRICTOR
- BOOMSLANG • COBRA • CORAL SNAKE
- EGG-EATING SNAKE • FER-DE-LANCE • GARTER SNAKE
- GRASS SNAKE • HOGNOSE SNAKE • KING SNAKE
- MAMBA • NIGHT ADDER • PIT VIPER • PUFF ADDER
- PYTHON • RACER • RAT SNAKE • RATTLESNAKE
- SEA SNAKE • SIDEWINDER • TREE SNAKE
- WATER SNAKE • WHIP SNAKE

SNAPPER

THE SNAPPER FAMILY is a diverse group of over 100 species of tropical reef fish. Snappers are important food fish and are often considered to be game fish, because they are a challenge for the sea angler to land. They are deep-bodied fish with a large head, somewhat flattened on top as it slopes up to meet the front of the dorsal fin. The mouth is large and the jaws, with their sharp teeth, slope down to the corners, giving the fish a disgruntled look.

The name is derived from the way the landed fish suddenly and very forcibly opens and shuts its jaws as it is dying, which sometimes causes bad wounds to the hands of unwary fishers sorting their catch. Snappers are usually less the 2 feet (60 cm) long, but the mangrove red snapper, *Lutjanus argentimaculatus*, attains 40 inches (1 m) in length. The dorsal fin is spiny in front but soft-rayed in the rear portion. The anal fin has several spines in front of its leading edge, and the pelvic fins are well forward, under the pectorals. Some species are gray to grayish green but many are beautifully colored: red, rose or in the case of the emperor red snapper, *L. sebae*, whitish with reddish brown bands. Yellow is another predominant color. The lane snapper, *L. synagris*, of Florida to Brazil is striped red and yellow, and the mutton snapper, *L. analis*, of the West Indies is banded yellow and green. Snappers are found in all warm seas. The greatest number of species are found throughout the Indonesian and Australian area, although many are also found in the tropical waters of the Atlantic coast of North America.

Small shoals

Some snappers move about in groups of less than a dozen, but other species form large aggregations. They occur particularly around coral reefs, usually at depths of 24–90 feet (7–27 m). They are also seen near mangrove swamps and docks, always ready to investigate a possible source of food. Snappers feed mainly at night. Their hunting method is to stalk living prey until a few feet from it, then to make a sudden dash, seize the prey, and swim leisurely back to the starting point.

Snapping up anything

Snappers are euryphagous, which means they refuse nothing edible. They feed mainly on fish, but they will take crabs, lobsters and prawns, barnacles, octopus and squid, brittlestars, sea squirts, pyrosoma, salps, sea butterflies, worms and mollusks. They also take a small amount of plant food. Snappers will also take scraps thrown overboard from ships, including vegetable waste.

The kind of food taken depends very much on which prey happens to be plentiful. There is also a change of diet with age, the young snappers feeding mainly on plankton or small fish. A study of the food taken by one species over time showed that of the total food taken, 62 percent was fish and 25 percent crustaceans. In another the diet was 49 percent fish, 12 percent crustaceans and 12 percent squid and plankton.

Breeding condition

In tropical waters the temperature varies very little throughout the year. Off East Africa, for example, temperatures vary from 24–29° C, and snappers spawn throughout the year. In Indian waters breeding seems to coincide with the cold season following the onset of the northeast monsoon in September and October. Off tropical America there seems to be two breeding seasons for snappers.

It is simple to note when a fish population is actively breeding by examining individuals caught in fishing nets The males, even without gutting, can be tested by gently pressing the flanks to see if milt (fluid containing the sperm) flows out. To find out where spawning takes place, to net the eggs and larvae and to follow the life history of the fish is far more difficult.

*A mixed-species group of yellow snapper (**Lutjanus apodus***) *and gray snapper (**L. griseus***) *shoals over a coral reef in the Cayman Islands in the Caribbean.*

The bluestripe snapper, Lutjanus kasmira, *has a wide distribution across the Pacific and Indian Oceans. The species is often found in large shoals such as this above coral reefs, both in shallow lagoons and on the outer slopes of reefs.*

Mature fish are not always in breeding condition. After spawning their reproductive organs are spent and the fish are in what is called a resting condition. After this the organs begin to increase in size again and the germ cells become active. A male fish not yet ready for breeding may give out milt, but when fully ready, it gives out milt when even a slight pressure is applied to the flanks. A fully ready female gives out eggs when her flanks are squeezed To detect other stages of breeding condition, she must be gutted and the size and maturity of her eggs examined.

Unpredictable poison

There is a kind of food poisoning in humans called ciguatera. which causes muscular pains, cramp, nausea, diarrhea, and even paralysis. Early symptoms are a tingling of the lips and throat and a sensation reversal in which hot substances in the mouth feel cold and cold things feel hot. Snappers are among the 300 or more species of food fish that can cause ciguatera. Ciguatera is sporadic and unpredictable and is probably caused by a fish eating a particular kind of blue-green alga, perhaps at a certain time of year or in a particular stage of the alga's growth. The poison persists and accumulates in the food chain. Snappers kept in aquariums for over a year can still cause ciguatera when they are fed to animals.

YELLOWTAIL SNAPPER

CLASS	**Osteichthyes**
ORDER	**Perciformes**
FAMILY	**Lutjanidae**
GENUS AND SPECIES	***Ocyurus chrysurus***

WEIGHT
Up to 8 lb. (3.7 kg)

LENGTH
Up to 33 in. (85 cm)

DISTINCTIVE FEATURES
Yellow mid-lateral stripe; yellow tail; pale yellow blotches above mid-body; background color violet to blue above, white below; red cast around mid-lateral stripe

DIET
Adults: plankton and bottom-dwelling animals, including fish, crustaceans, worms, gastropods (marine snails), and cephalopods (squid and their relatives); juveniles: plankton

BREEDING
Breeding season: spawns throughout year; eggs probably pelagic

LIFE SPAN
Up to 14 years

HABITAT
Coastal waters, mostly coral reefs (adults) and weed beds (juveniles), down to 590 ft. (180 m). Usually found in midwater, not close to bottom.

DISTRIBUTION
Coastal waters of Caribbean and western Atlantic from Massachussetts to southeastern Brazil. Most common in Bahamas and southern Florida.

STATUS
Common

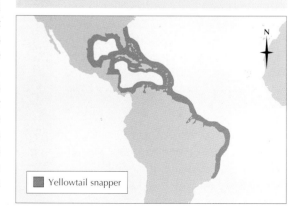

Yellowtail snapper

SNAPPING TURTLE

THE SNAPPING TURTLES OF North and South America are named for their powerful bite and need to be handled with caution. A full-grown snapping turtle can easily break a pencil in two or severely maul a person's hand. Snapping turtles are heavily built, with large heads and strong, hooked jaws; they are unable to retract their limbs into the shell. The common snapping turtle, *Chelydra serpentina*, usually known simply as the snapping turtle, has a shell length of up to 15 inches (38 cm) but it is proportionately very heavy and may weigh up to 50 pounds (23 kg). It is drab in color, with a gray, black or brown top. The tail is half the length of the shell and bears a row of scales like a crest on the upper surface. The feet are partly webbed and bear strong claws. The skin is greenish and the shell is often covered with green algae. The plastron, the underside of the shell, is reduced in size and forms a crosslike shape, with the turtle's limbs fitting between the arms of the cross.

The common snapping turtle is the most widespread turtle in North America, ranging from southern Canada to Florida and southern Texas, and in parts of Mexico and south to Ecuador. The other species of snapping turtle, the alligator snapper, *Macroclemys temmincki*, is restricted to the United States, where it is found in the Mississippi basin, from Kansas, Iowa and Illinois eastward as far as Georgia and northern Florida. It is also found in parts of Texas. It is one of the largest freshwater turtles and can grow to 200 pounds (20 kg). The shell may be nearly 2½ feet (75 cm) in length; it is extremely rough and features three ridges. The eyes of the alligator snapper are on the side of its head, unlike those of the common snapping turtle.

Snapping turtles have a more aquatic lifestyle than most other freshwater turtles and spend most of their lives in muddy ponds, lakes and rivers. Snapping turtles are usually only aggressive when they are encountered on land. They hibernate but can sometimes be seen swimming in lakes under ice during the winter months.

Active and passive hunters

The common snapping turtle actively hunts for its food, which consists of plants, carrion, insects, fish, frogs, ducklings and young muskrats. Live prey is caught with a quick thrust of the head and snap of the jaws and is then pulled apart by the mouth and claws. The fish-eating habits of snapping turtles often bring them into conflict with anglers or the owners of fish farms. Too many snapping turtles in a fish pond can lead to too few fish, but some studies have suggested that the harmful effect of snapping turtles on fish populations is often exaggerated. In many places, however, snapping turtles are trapped either because of their supposed depredations or so that they may be used for snapper soup.

Unlike the common snapping turtle, the alligator snapper is a passive hunter. It lies in wait for its prey, half-buried in the mud and camouflaged by the algae growing on its shell which, combined with its rough texture and ability to remain motionless for some time, gives the shell a rocklike appearance. From this concealed position the alligator snapper lures small fish into its mouth using a remarkable piece of deception.

This adult male snapping turtle is making a threat display with its powerful hooked jaws.

SNAPPING TURTLES

CLASS	**Reptilia**
ORDER	**Testudines**
FAMILY	**Chelydridae**

GENUS AND SPECIES **Snapping turtle,**
***Chelydra serpentina*; alligator snapper,**
Macroclemys temmincki

ALTERNATIVE NAMES
**Common snapping turtle; snapper
(*C. serpentina* only)**

LENGTH
***C. serpentina* shell length: up to 15 in.
(38 cm); *M. temmincki* shell length: up to
30 in. (75 cm)**

DISTINCTIVE FEATURES
**Large head; long tails; powerful bite. *C.
serpentina*: sawtooth projections on dorsal
surface of tail. *M. temmincki*: smooth tail;
three prominent ridges along back of shell.**

DIET
***C. serpentina*: nearly all animal and plant
material, living or dead; *M. temmincki*:
variety of foods, but not usually plants**

BREEDING
**Breeding season: eggs laid during summer;
hatching period: about 4 months, but
C. serpentina eggs may overwinter**

LIFE SPAN
Not known

HABITAT
***C. serpentina*: almost all freshwater habitats;
also in slightly salty water near estuaries and
the sea. *M. temmincki*: mostly deep rivers
and lakes with muddy bottoms.**

DISTRIBUTION
***C. serpentina*: southern Canada south to
Ecuador; *M. temmincki*: U.S., in Mississippi
basin and surrounding areas**

STATUS
Both species common

*Common snapping
turtles are poor
swimmers and usually
prefer to walk across
the bottoms of rivers
and lakes rather
than swimming.*

The tongue is forked and the two branches are
fat and wormlike; moreover, when filled with
blood, the tongue becomes bright pink. The
turtle moves its tongue, making the wormlike
tips wriggle, and this action attracts small fish,
which are then promptly snapped up. Larger
prey, including ducklings, are also caught, but by
more active hunting, not by means of the lure.

Alligator snappers are virtually omnivorous.
As well as taking fish, amphibians and reptiles,
including other turtles, they raid henhouses in
search of eggs, chicks and even adult birds. They
also eat carrion, but do not usually eat plants.

Overwintering eggs

Common snapping turtles may crawl some
distance from water to find a suitable place to
make a nest. They seem to prefer open areas and
even cultivated land. The nest is dug with the
hind feet. The usual size of a clutch is 20 to 30
eggs, but there is one report of a very large
female producing 83. Hatching takes place
during the late summer, and eggs that are laid
late may not hatch until the following spring.

As with many species of turtles, the sex of
the hatchlings, is determined by the temperature
at which the eggs have developed. If the average
temperature is above about 86° F (30° C) or
below 68° F (20° C), they will be females. If it is
between these temperatures, they will be males.
Breeding habits of the alligator snapper are
similar but it lays its eggs, which may number
from 17 to 44, nearer to the water.

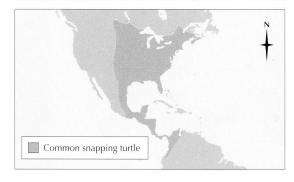

Common snapping turtle

SNIPE

THE SNIPES AND THEIR close relatives, the woodcocks, belong to the family Scolopacidae. Their general form is like that of other shorebirds, but their legs are not particularly long and their necks are fairly short. Their bills are very long and straight and their eyes are set well back on their heads, giving the snipes a characteristic appearance. The plumage is generally dark brown, mottled and barred.

There are 18 species of snipes in 3 genera: *Gallinago* (15 species), *Lymnocryptes* (1 species) and *Coenocorypha* (2 species). The 6 species of woodcocks, which are all in the genus *Scolopax*, are discussed elsewhere.

Species look alike

The snipes have similarly patterned plumages that provide them with excellent camouflage. The common snipe, *Gallinago gallinago*, is 10–10⅗ inches (25–27 cm) long with a bill of 2⅗–2¾ inches (6–7 cm). It is cosmopolitan, breeding in much of northern North America, Europe and Asia and wintering as far south as northern South America

and parts of Africa and southern Asia. Its brown plumage is barred with black to give an effect of horizontal stripes, the crown is black and there are black stripes through the eyes. The underparts are whitish. The subspecies of common snipe found in the Americas, which is often known as Wilson's snipe, is now considered to be a full species by some authorities.

The great snipe, *G. media*, is larger than the common snipe but with a proportionately shorter bill. It breeds in Scandinavia, eastern Europe and western Asia. In central and eastern Siberia the great snipe is replaced by Swinhoe's snipe, *G. megala*, which migrates south as far as Southeast Asia and India in winter. The smallest species of snipe is the jack snipe, *Lymnocryptes minimus*, which is only 6⅗–7½ inches (17–19 cm) long with a 1½-inch (4-cm) bill. Its plumage is mottled rather than barred. The jack snipe breeds in northern Eurasia, but like many snipes the true habits of this species are not fully understood because of the bird's skulking behavior, which makes observation difficult.

The several South American snipes are much less well known than their northern relatives. Pictured are two Magellan snipes, Gallinago paraguaiae, *on the Falkland Islands to the east of Argentina.*

The long bills of snipes (common snipe, above) have flexible tips, enabling the birds to snap up food while the bill-tips are buried in mud or damp earth.

Undercover birds

Snipes live in open country, usually in wet places such as marshes, damp meadows and moors or in the swampy arctic tundra. Their mottled plumage makes them very difficult to find among the low vegetation unless they are flushed, when they fly up on rapidly whirring wings and zigzag before shooting away. The alarm call given by common snipes as they flee is a harsh grating or a rapid squeaking: *ship-per chip-per*. Snipes keep to the cover of vegetation and are most active in the evening or early morning, as is suggested by their large eyes.

Flexible bills

Snipes feed in the mud bordering lakes, rivers and marshes or in the damp ground of fields and swamps, where they probe for a wide variety of small animals. Their food is mainly worms and insects, but they also eat snails, small crustaceans, spiders and wood lice. Up to 80 percent of their diet consists of the larval stages of aquatic insects. Snipes also eat smaller quantities of seeds and plant stems.

Like a woodcock, a snipe can easily open its bill underground because the tips of the mandibles are very flexible and can be forced apart even when the bill is buried to its base in the soil. Smaller animals can be eaten in this way without the bill being withdrawn.

COMMON SNIPE

CLASS	**Aves**
ORDER	**Charadriiformes**
FAMILY	**Scolopacidae**
GENUS AND SPECIES	*Gallinago gallinago*

ALTERNATIVE NAMES
Wilson's snipe (Americas only); fantail snipe

WEIGHT
2⅔–6⅓ oz. (75–180 g)

LENGTH
Head to tail: 10–10⅔ in. (25–27 cm); wingspan: 17⅓–18½ in. (44–47 cm)

DISTINCTIVE FEATURES
Extremely long, straight bill; stocky body with bull neck; proportionately short legs; long toes; brown plumage with complex pattern of bars and streaks in black, buff, chestnut and white; pure white belly

DIET
Mainly aquatic insects, earthworms, small crustaceans, small snails and spiders; some plant stems and seeds

BREEDING
Breeding season: eggs laid April–June; number of eggs: usually 4; incubation period: 17–20 days; fledging period: 19–20 days; breeding interval: 1 year

LIFE SPAN
Up to 12 years

HABITAT
Marshes, damp tussocky vegetation, marshy tundra, muddy lake shores and rice fields

DISTRIBUTION
Breeds throughout northern half of North America, Europe and Asia; winters as far south as Colombia, Kenya and India

STATUS
Common

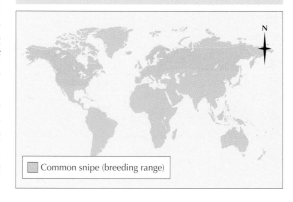

Common snipe (breeding range)

Hidden chicks

With the exception of the Chatham Island snipe, *Coenocorypha pusilla*, which nests among the roots and buttresses of trees and under fallen logs, snipes nest in grass-lined hollows inside tussocks or clumps of vegetation.

In the common snipe the clutch consists of four eggs, sometimes three or five, which are usually olive brown with dark blotches. The female incubates the eggs for 17–20 days and the chicks leave the nest shortly after their down has dried. After the chicks have hatched, if disturbed by a predator, the adults run conspicuously to and fro or perform distraction displays. The chicks are, however, inconspicuous, their down being brown and black, like a tortoiseshell domestic cat. When disturbed the chicks burrow headfirst into the coarse grass, which effectively hides their outlines. The chicks are fed by both parents and fly when 19–20 days old.

Drumming displays

The courtship displays of snipes are known as drumming or beating flights. A drumming snipe flies to a considerable height and then dives rapidly with tail spread and the wings half-closed and beating slowly. This is accompanied by a soft, resonant drumming, which is repeated as the snipe glides and soars. The drumming is produced by air hitting the very rigid outer tail feathers, the vanes of which are held tightly together by an unusually large number of hooklets. The feathers vibrate rapidly to produce a humming note, and the tremulous effect is added by the slipstream of air being pulsed by the slowly beating wings. The pintail snipe, *Gallinago stenura*, has 26 tail feathers, 12 more than the common snipe, and the 8 outer pairs are stiffened for drumming.

Communal displays

The great snipe differs from the others in having communal displays called leks, rather like those of the ruff, *Philomachus pugnax*, discussed elsewhere. Large numbers of male great snipes gather at a particular spot in the evening and display throughout the night, twittering in chorus, snapping their bills or slowly flapping their wings. The little known noble snipe, of South America, may also display at leks.

Several snipes have specially adapted tail feathers that produce a soft drumming sound as the birds perform their display flights.

SNOEK

THE SNOEK IS RELATED TO the castor-oil fish, *Ruvettus pretiosus*, and the snake mackerel, *Gempylus serpens*, and is an important food fish in the Southern Hemisphere. It is like the snake mackerel in shape. Its back is bluish black, the rest of the body is silvery with faint dark bands across the flanks. The eye is golden.

Snoek is the Dutch name for a freshwater pike and was used by the early Dutch settlers in South Africa. In Australia and New Zealand the fish is known as barracouta, due to its resemblance to the predatory barracuda. It is also sometimes called snook in Australia, and sometimes pike or sea pike. There are other confusing uses of these names, and additionally there is a snook, *Centropomus parallelus*, in tropical American waters, which belongs to a different family, the Centropomidae.

The single species of snoek lives in waters of 50–68° F (10–20° C), from the surface to a depth of 240 feet (73 m). It is distributed from the seas around South Africa to Tristan da Cunha, Argentina, Chile, New Zealand and Australia.

Fish fanciers

Snoek swim in large shoals. They are active fish that prey on other fish. Off South Africa they feed especially on the sardine, *Sardinops sagax*, but any small fish living near the surface are taken, including the young of their own species. They are also important predators of the South African pilchard *Sardinops ocellatus* and the maasbanker, or South African mackerel, *Trachurus trachurus*. Snoek are described as singling out individual fish, then pursuing them for as much as half a minute, using their greater stamina to tire the smaller prey, which can outmaneuver the

snoek to escape. Sardines will often leap 2–3 feet (60–90 cm) out of the water when pursued by a snoek, which keeps up the chase when both have fallen back into the water. Sometimes hundreds of sardines, each followed by a snoek, can be seen, all leaping out of the sea together. On occasions both prey and predator have leaped into a boat. Snoek also commonly hunt pelagic (oceanic) crustaceans such as krill, of the genera *Ephausia* and *Nyctiphanes*.

Seasonal appetites

Although voracious feeders that grow very fat during their feeding season, snoek, like so many other species, stop feeding as the breeding season approaches. The breeding season off South Africa is spring, but it varies in other parts of the snoek's range. Some of the spawning grounds have been located, largely by examining the stomachs of predatory fish containing snoek fry. Studies of the snoek have revealed details of its life history. Roes from the ripe snoek were taken and their contents, eggs and sperm, were released into a seawater aquarium onboard ship. The artificially fertilized eggs, $\frac{1}{25}$ inch in diameter, floated at the surface because of the oil droplet in each of them. They hatched in 48 hours, and the newly hatched larvae, $\frac{1}{12}$ inch long, floated upside down at the surface, buoyed up by the oil droplet and the remains of the yolk sac. By the end of 9 days the larvae were $\frac{1}{6}$ inch long, after which time they died. The difficulty was in finding food for them. They refused diatoms and brine shrimp larvae. Extract of beef liver, which was accepted by the larvae, fouled the water. The larvae were finally fed on drops of human blood by the devoted scientists.

When hooked, the snoek is likely to bleed and lose many scales. Fishers must take care if they want to set the animal free after capture, or the released fish will not survive.

SNOEK

CLASS	**Osteichthyes**
ORDER	**Perciformes**
FAMILY	**Gempylidae**
GENUS AND SPECIES	***Thrysites atun***

ALTERNATIVE NAMES
Barracouta; sierra; couta; sea pike

WEIGHT
Up to 13 lb. (6 kg)

LENGTH
Up to 80 in. (2 m)

DISTINCTIVE FEATURES
Large and elongated; long jaws; strong front canine teeth; small smooth scales and small forked spines covering body; few finlets following dorsal and anal fins

DIET
Cephalopods (squid and their relatives); fish; pelagic (oceanic) crustaceans

BREEDING
Age at first breeding: 2–4 years; breeding season: spawns in spring; eggs planktonic

LIFE SPAN
Up to 10 years

HABITAT
Continental shelves or around islands; schools near the bottom or midwater, or surface at night

DISTRIBUTION
Southern Pacific, Indian and Atlantic Oceans

STATUS
Abundant

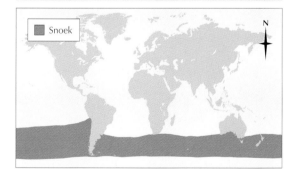

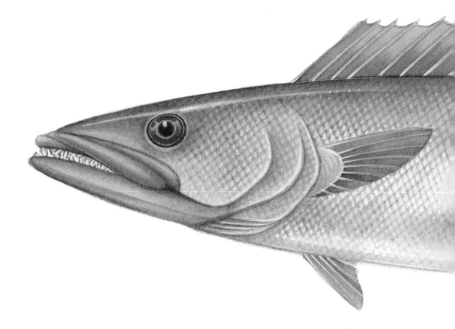

Huge catches

Snoek are caught by various forms of line fishing such as trolling and jigging. If there is a local abundance of prey fish, the snoek ignore the bait, but otherwise they will take any flesh from fish and squid to offal. The catches vary widely from year to year for reasons that are not fully clear. It may be that there are 7–9-year cycles of abundance and scarcity. On the other hand, snoek are mainly fished from small vessels, which are less numerous in bad weather, so the cycles could be due only to changes in the weather conditions. Although the snoek is an important food fish, the total catches for South Africa have varied from 4 million to 9 million pounds (1,800–4,100 tonnes) per year. The returns for Australia are smaller, with an average catch of 4 million pounds (1,800 tonnes). The fish are eaten fresh, salted or smoked.

Tiger of the sea

Shortly after World War II, when Britain was desperately short of food, quantities of snoek were sent from South Africa. Instead of meeting with gratitude, the snoek became a comedians' joke and material for the cartoonist. One cartoon showed a housewife preparing to open a can of snoek while her husband stood by with an ax poised to deal with "the tiger of the sea." Perhaps the fish was not popular in Britain. This contrasts with the account given in a report from the *Discovery II*. In November 1933, the ship lay at anchor off Tristan da Cunha. Snoek were lingering around the gangway lights at night, so the scientists aboard started to fish for them, landing ¾ ton of them in 3 hours. "The fish were placed in the ship's cold store as soon as we finished cleaning them next morning," the report reads, "and provided at least one course per day for all who cared for them until we reached New Zealand some 2½ months later."

Possessing similar sleek lines to the predatory barracuda, the snoek is also a voracious feeder, and can inflict a lot of damage on human handlers with its large canine teeth.

SNOW LEOPARD

The snow leopard's range is limited to the snowy mountains of central Asia. In the winter it may descend to lower, warmer altitudes, where prey is more plentiful.

THE SNOW LEOPARD, OR ounce, grows to a length of about 7⅓ feet (2.2 m) or more, 40 inches (1 m) of which is accounted for by the animal's tail. It is identified by its long, thick, almost woolly fur, the ground color of which is pale whitish gray to bronzy olive; the underparts are often pure white. There is sometimes a black streak down the midline of the animal's back, and the rest of the coat is spotted with large irregular rosettes. It also has thick tufts of fur on its paws to keep them warm.

The snow leopard is much smaller than other big cats, with an average male weighing only about 110 pounds (50 kg). It has the typical physical features of big cats, such as retractable claws, large canines with which to grasp and throttle prey, a rough tongue to remove the prey's fur and skin, and forward-facing eyes that provide it with excellent night vision.

Equipped for mountain dwelling

Snow leopards occur in the mountain ranges of central southern Russia, such as the Pamirs, and from there eastward to Tibet and the Himalayas. Farther north they extend into the Altai and Sayan Mountains, into Mongolia and into western China.

Probably the least understood of the big cats, the snow leopard inhabits sparsely populated areas in its range. It is a solitary animal, and one pair of snow leopards may occupy the whole of a large valley. It lives at fairly high altitudes, often encountering cool and even very cold conditions. Accounts differ as to exactly how far the snow leopard travels up into mountainous areas. However, it is likely that the animal typically ranges between altitudes of 1,980–16,500 feet (600–5,000 m), with occasional forays to heights of perhaps 19,800 feet (6,000 m).

The snow leopard migrates with the seasons, coming down to lower levels in the winter to escape the harshness of the season's snows and storms. However, it is adapted to resist the cold that it does encounter. Its thick, luxuriant fur provides excellent insulation, and its small ears lose little heat through radiation. Even its thickly furred tail is useful in this respect: the cat curls it around itself when sleeping, usually positioning it as a muffler to protect the bare tip of its nose.

Camouflaged for hunting

The snow leopard's pale coat helps it blend in with the rocks and snow of high mountains. Moreover, the irregular patterning of its rosettes

SNOW LEOPARD

CLASS	**Mammalia**
ORDER	**Carnivora**
FAMILY	**Felidae**
GENUS AND SPECIES	***Panthera uncia***

ALTERNATIVE NAMES
Ounce; *Xue bao* (China); *barfani chita* (Pakistan); *sarken* (Tibet)

WEIGHT
77–121 lb. (35–55 kg)

LENGTH
Head and body: 40–48 in. (1–1.2 m); shoulder height: 24 in. (60 cm); tail: about 40 in. (1 m)

DISTINCTIVE FEATURES
Creamy, long fur, paler on undersides and chin, with dark rings and rosettes; long, well-furred tail

DIET
Small mammals; birds; large ungulates

BREEDING
Age at first breeding: 2–3 years; breeding season: January–March; number of young: usually 2 or 3; gestation period: 98–104 days; breeding interval: 2 years

LIFE SPAN
Up to 21 years

HABITAT
Montane plateaus; cold, arid lands; steppes, grassland and steep, rocky slopes at high altitude 1,980–16,500 ft. (600–5,000 m)

DISTRIBUTION
Afghanistan; Pakistan; northern India; Nepal; Butan; Tibet; China; Mongolia; Kazakhstan; Kyrgyzstan; Russia; Tajikistan; Uzbekistan. Scattered small populations.

STATUS
Endangered; estimated population: 4,000 to 7,000

Snow leopard

help it to merge with the dappled shadows of juniper bushes and the spruce and birch trees that grow below the tree line of its range.

Snow leopards hunt by day and night, taking both small creatures, such as ground squirrels and pikas (relatives of rabbits), and the larger ibex, tahr, marhor and gazelle. They also kill domestic animals when they are available, such as sheep, goats and cattle, especially in winter when other prey is scarce. Snow leopards ambush their prey or stealthily creep as close as possible and then pounce on the unsuspecting victim. The last few yards are often covered in a single leap, as these big cats are capable of covering 15 yards (13.5 m) in one bound.

The snow leopard's lengthy tail helps the animal to maintain its balance on the steep mountainsides of its native habitat.

Life cycle adapted to climate

To avoid having the offspring born into icy, wintery conditions, snow leopards have a fixed breeding season. They do not mate until late winter or early spring, and after a gestation of 90–100 days, two to four cubs are born in April. At this time the days become warmer and hunting is consequently easier, with much breeding activity taking place near suitable prey in the mountains. The cubs are born in a cave or a rocky cleft and begin to follow their mother on her hunting trips in July. They remain with her through their first winter at least, before departing the following year.

Slim survival chances

Reliable data is scarce, but the snow leopard seems to be very rare throughout its range. Naturalists estimate that perhaps only 4,000 to 7,000 exist in the wild; about 600 survive in zoos in the United States. The decline in the animal's population has partly been caused by habitat loss. For example, in Nepal an increasing amount of alpine land is being used for pasture. However, the main reason for the falling numbers is the fact that skin traders hunt the animal for its valuable fur, which is made into pelts, and its bones, which are used in traditional Asian medicines.

International and local laws protect snow leopards, and because of the international agreements restricting trade in the furs of endangered cat species, the value of snow leopard pelt has fallen drastically. However, the pelts are still sold widely in the illegal skin trade. Moreover, snow leopards are regarded as pests in many areas, as they threaten local livestock. Most affected are the inhabitants of remote villages where survival from year to year is a struggle and where every

animal is a potential source of income. Native peoples set poison-tipped spears in the ground on trails favored by snow leopards, with each spear sticking out at an angle. The poison is very potent: a fully grown wild blue sheep runs only 30 yards (27 m) or so before it collapses dead when wounded by such a spear.

Survival strategies

The only way the snow leopard will survive is by the creation of large and defensible sanctuaries and the development of ecologically sound economies for the remote villages within its range. Recently, programs have been instigated that seek to provide incentives, in the form of food and clothing, to encourage native peoples to cease hunting the snow leopard. The organizations involved are looking to reward local peoples for protecting the animal rather than competing with it. These moves have been spearheaded by the International Snow Leopard Trust.

In another innovative step, selective breeding programs have been set up in the United States to create and maintain a stable captive population and minimize the effects of inbreeding, which can be detrimental to a species. However, it may already be too late to ensure the survival of this rare and striking animal.

Snow leopard cubs are born blind. They are raised by their mother alone, and remain with her throughout their first winter. They are fully grown at 4 years.

SNOW PETREL

THE SNOW PETREL, and its close relatives the Antarctic petrel, *Thalassoica antartica*, and the Antarctic or southern fulmar, *Fulmarus glacialoides*, nest farther south than any other birds. They have been found nesting in Antarctic mountain ranges many miles from open water.

The snow petrel has a pure white plumage, which is offset by its black eyes, bill and legs. The whiteness of the bird's plumage makes it very difficult to see against the snow of its Antarctic habitat, as compared with a polar bear of the Arctic, for instance, the fur of which is nearer a dirty yellow color. By contrast, the Antarctic petrel, which has a similar distribution, has a brown-and-white checkered plumage.

Snow petrels are 12–16 inches (30–40 cm) long and, like the fulmar of the North Atlantic, *F. glacialis*, have a typical petrel bill with the characteristic tubular nostrils. However, they are placed in a genus of their own. They breed around the coast of the Antarctic continent, inland to about 195 miles (325 km), and on the islands of the Scotia Sea as far north as South Georgia Island.

Birds of the pack ice

The life of the snow petrels is very closely linked to the drifting pack ice of the Southern Ocean. The birds fly low over the waves in the same manner as other petrels, whereas in the pack ice itself they sit in tight groups on floes, abruptly taking off into flight together as if prompted by a signal. The petrels' dependence on pack ice explains their distribution, as the current from the Weddell Sea carries ice northward around South Georgia Island. However, snow petrels do not breed on Heard Island, which is located at about the same latitude as South Georgia Island. The only snow petrel to have been seen on Heard Island arrived when the pack ice came unusually close to it.

Snow petrels occur mainly in areas with 10–50 percent ice cover. They rarely alight on the water, preferring to rest on icebergs or ice floes. They prefer to ride storms out on the wing, flying in the wave troughs for shelter.

Keen noses

Like other petrels, snow petrels feed on small animals living near the surface of the sea. Much of their feeding takes place at night, and in the pack ice they fish in the leads (water channels) and the pools of open water that run between the ice floes. Most of the snow petrels' food consists of small crustaceans, including krill and

amphipods, but they also catch fish. Snow petrels also feed on carrion, for example whale blubber, seal carcasses and placentae, dead seabirds and excreta. While they are rearing their chicks they catch the immature stages of such fish as Antarctic cod, which live at the surface.

In general, birds have a poorly developed sense of smell but that of the petrel, with its tubular nostrils, is better developed than most. Moreover, the snow petrel also has larger olfactory organs than any other petrel that naturalists have examined. This would suggest that the snow petrel finds its food by smell, although it is not clear how the sense of smell can be of use in catching small animals by plunge-diving.

Snow butterflies

Snow petrels breed on sea cliffs or inland mountain faces, nesting on ledges or between boulders. In the northern parts of their range, snow petrels fly around their nesting areas at any time of the year, and even farther south snow petrels are still visible when other petrels have dispersed out to sea for the winter. Throughout the time that the snow petrels are still visible around the cliffs they call, particularly at night. They have two calls, a raucous shrieking and a caw repeated four to five times. There is also an aerial display called the butterfly flight, in which

The snow petrel builds its nest in rock crevices. It is the only relative of the fulmars to build its nest under cover.

The snow petrel catches its prey, such as fish and krill, when flying over the surface of the sea in areas near pack ice.

SNOW PETREL

CLASS	**Aves**
ORDER	**Procellariiformes**
FAMILY	**Procellariidae**
GENUS AND SPECIES	***Pagodroma nivea***

ALTERNATIVE NAMES
Gray-faced petrel; long-winged petrel

WEIGHT
About 9½ oz. (268 g)

LENGTH
Head to tail: about 12–16 in. (30–40 cm); wingspan: 2½–3 ft. (75–90 cm)

DISTINCTIVE FEATURES
All-white plumage; very small, black bill; black eyes and legs

DIET
Krill, fish, squid and carrion

BREEDING
Age at first breeding: probably 1 year; breeding season: eggs laid mostly November–December; number of eggs: 1; incubation period: 41–49 days; fledging period: 41–54 days; breeding interval: 1 year

LIFE SPAN
Usually 14–20 years

HABITAT
Cold seas with 10–50 percent pack ice; breeds on cliffs and rock faces up to 7,920 ft. (2,400 m) high and up to 195 mi. (325 km) inland

DISTRIBUTION
Antarctica and Antarctic islands; South Sandwich Islands; Scotia Sea (South Georgia Island to South Shetlands and Antarctic Peninsula, south to Palmer Archipelago)

STATUS
Locally common

a snow petrel flies on rapidly beating, almost quivering wings. The butterfly flight is thought to have some significance in courtship, but since it is even performed by immature birds its function is obscure.

Just before the single white egg is laid the snow petrels disappear from the cliffs. They head out to sea to feed and to accumulate enough food reserves for the female to form the egg and for the male to survive the first long spell of incubation. Incubation lasts 6–7 weeks and for a few days after the egg has hatched the fluffy chick is brooded. Then it is left alone, except when the parents return at night to feed it. The chick is able to fly at 41–54 days.

The population of snow petrels is stable, and may even be increasing. There are an estimated 2 million birds in various colonies distributed around the Ross Sea.

Nesting farthest south

For some years ornithologists have been aware that snow petrels have nested in the Tottanfjella, a range of mountains 300 miles (480 km) east of the Weddell Sea. This was the farthest south that any bird was known to nest, but more significant than this is the distance of the nests from open water. To get food for their chicks the snow petrels have to make a round-trip of at least 600 miles (960 km), often against strong winds. Consequently, naturalists regard the snow petrels as having the most southerly nesting place of any bird species.

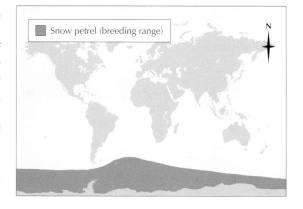

Snow petrel (breeding range)

SNOWY OWL

ALTHOUGH IT IS ONE OF the most powerful owls, the snowy owl is not the largest. It is 21–26 inches (53–66 cm) long, making it smaller than the eagle owl, *Bubo bubo*, of Europe and Asia. It has a similar appearance to a large barn owl, *Tyto alba*, but its plumage is whiter. The male snowy owl may be almost completely white or may have some dark barring and spots. The female is larger than the male and her plumage is dark brown or black and more extensively barred on the head, back, wings and underparts. Dark males may, however, be darker than light females. Juveniles have a similar coloration to the female. The feathers extend down the legs and may even cover the claws. The bill is black, but almost obscured by feathers.

The snowy owl has large eyes, which are usually a feature of nocturnal animals. The bird often hunts during the light nights of the Arctic in summer, although it is also active during the short daylight hours in winter.

Snowy owls breed in the tundra of the Old and New Worlds. In America they range from the far north of Greenland to the southern shores of Hudson Bay; in the Old World they breed in the north of Siberia, in northern Scandinavia, on the islands of Spitsbergen, Novaya Zemlya and in Iceland.

A new nesting ground

Snowy owls inhabit the barren, treeless wastes of the north, extending for as far north as there is snow-free ground in the summer. During the winter there is a general movement of the birds southward, and this is quite spectacular in years when there is a shortage of food in the tundra. In such years snowy owls are found as far south as California, Bermuda, England and France, becoming spectacular additions to the local bird life.

In 1967 the snowy owl's breeding range took a well-publicized extension southward when a nest was discovered on Fetlar, one of the Shetland Islands, to the north of Scotland. For a few years before this discovery, snowy owls had often been seen on Fetlar and neighboring islands and had probably been induced to stay by the plentiful supply of rabbits and the similarity of the terrain to the Arctic tundra; the nesting area on Fetlar is rocky with low herbage. The pair continued to breed successfully in the following years. It is possible that snowy owls used to breed in the Shetland Islands as they were once common there before collecting took its toll on their population, but there are no clear records of nesting. However, colonization in the Shetlands was not to be long-lasting.

The snowy owl's primarily white plumage offers the bird highly effective camouflage in the snowy landscapes of the Arctic.

With its rounded head and broad wings, the snowy owl is the most heavily built large white Arctic bird.

SNOWY OWL

CLASS	**Aves**
ORDER	**Strigiformes**
FAMILY	**Strigidae**
GENUS AND SPECIES	*Nyctea scandiaca*

WEIGHT
2⅔–6½ lb. (1.2–2.9 kg)

LENGTH
Head to tail: 21–26 in. (53–66 cm)

DISTINCTIVE FEATURES
Large, earless owl; fully feathered legs and toes. Adult male: white with a few black spots. Female and juvenile: heavily barred black or dark brownish plumage, appearing gray at distance.

DIET
Lemmings and voles on tundra; wider variety of mammals and birds elsewhere

BREEDING
Age at first breeding: 3–4 years; breeding season: eggs laid in May; number of eggs: 3 to 9; incubation period: 30–33 days; fledging period: 43–50 days; breeding interval: 1 year if food is available, but often 2 years or more

LIFE SPAN
Up to 17 years

HABITAT
Breeding: Arctic tundra, usually at low altitudes. Winter: any open landscapes.

DISTRIBUTION
Holarctic (northern parts of the Old and New Worlds). Breeding: Arctic Finland and Scandinavia; Russia; Siberia; Alaska; Canada; Greenland. Winter: farther south, range varies from year to year.

STATUS
Scarce

Nomadic when prey is scarce

Snowy owls prey on a wide variety of small animals, such as lemmings, mice, rats, moles, shrews rabbits, hares and ground squirrels, as well as on many birds, including oystercatchers, Arctic skuas, eider ducks, gulls and buntings. If none of these prey species are available, they may fish, take animals and birds caught in traps, steal poultry or forage for insects. The owls hunt from a perch on a rock or post, or quarter the ground on long glides with deep wing beats, almost like a hawk. Birds are sometimes caught in the air.

In many parts of its range, the chief food of snowy owls is lemmings (discussed elsewhere in this encyclopedia), although on Fetlar, for instance, they feed mainly on rabbits. Lemming populations undergo vast cyclic changes about every 4 years. These changes have profound repercussions on their predators, such as the snowy owl, the Arctic fox and the pomarine skua. Although snowy owls migrate farther south in years of lemming scarcity, they may, like the skuas, completely fail to breed. However, when lemmings are abundant the owls may lay more eggs than usual.

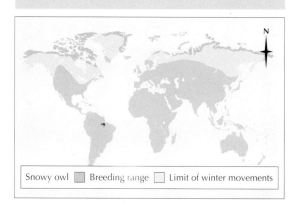

Snowy owl ▢ Breeding range ▢ Limit of winter movements

Staggered hatching

Snowy owls make no nest but lay their eggs in a depression in the ground on a knoll or hillock that gives the sitting bird a good view of the surrounding country. Consequently, it is very difficult to approach a snowy owl's nest without disturbing the sitting bird. Even if it is possible to approach without being seen from the nest, the male will give the alarm from its lookout nearby. The alarm call is a harsh bark, repeated up to six times, and if a human intruder approaches the nest, the owls may dive at the newcomer or show a threat display in which the snowy owl lowers its body with feathers fluffed, wings spread and eyes glaring. Faced with an attack from a skua, a snowy owl may thrust its wings forward and flap them as a form of defense. Large birds such as ravens and greater black-backed gulls are chased away from the nesting area.

The number of eggs in a clutch is usually 5 to 8, but there may be as few as 3 or as many as 14, depending on the abundance of food. The female incubates the eggs for 30–33 days while the male stands guard and brings her food. Incubation starts as soon as the first egg is laid, so the chicks hatch one after another and there may be a considerable difference in size between the oldest and the youngest. The female feeds the chicks, which beg for food by nibbling at her bill and feathers. As well as the threat of heavy rains, the eggs and young are preyed upon by a variety of animals, including Arctic foxes, jaegers, skuas and huskies. As the chicks get older, the male also feeds them, instead of just passing the prey to the female, and the female goes foraging for food herself. The chicks leave the nest when they are nearly 3 weeks old and shelter under rocks until they can fly, at about 6–7 weeks old. They stay together and continue to be fed by the parents for some time.

Snowy but camouflaged

Several animals that live in polar regions, such as polar bears, snow petrels and Arctic foxes, have white fur or feathers, and this is related to the nature of the predominantly white environment in which they live. Snow petrels and polar bears are usually found among pack ice, and Arctic foxes and snowshoe hares have white fur only during the winter. For the rest of the year, when the snow has gone, they have dark fur, which also blends against the background.

The white plumage of snowy owls seems to offer no camouflage at all and appears to make the birds highly conspicuous to both prey and predators. It apparently contradicts the effective camouflage that is a characteristic of most ground-nesting birds. However, only the female snowy owl sits on the nest, and she is not such a brilliant white color as the male. Moreover, her barred plumage makes her inconspicuous against the lichens and grasses of the tundra.

Female snowy owls incubate their eggs for about a month. Although up to nine young may be born, only four or five are likely to survive.

SOFT CORAL

This soft coral, genus Siphonogorgia, is a branching form that lives around Papua New Guinea. It can grow on vertical walls on the seaward side of fringing coral reefs.

SOFT CORALS ARE MEMBERS of a group of *octocorals*, so called because they have polyps divided radially into eight sections. Other octocorallian groups, to which soft corals are closely related, include the sea pens, sea fans (gorgonians) and organ pipe corals. These are all part of a larger group called the Cnidaria, which includes hard corals, jellyfish, anemones and sea firs. Cnidarians are all composed of single or group-living polyps with stinging cells. The soft corals, of which there are around 800 known species, are at their most diverse and abundant in tropical waters, in the Indian and Pacific Oceans and the Red Sea, although there are cold water examples, such as dead man's fingers, *Alcyonium digitatum*, which is described in a separate article. In the Caribbean, gorgonians dominate instead, and there are few soft corals.

Soft corals are varied in color and form, but their basic construction is constant. They are group-living organisms whose surface is covered by numerous relatively small polyps, usually less than ½ inch (1 cm) across, embedded in a soft material. The polyps are all interconnected by a series of vessels, and can share resources such as food and oxygen. In most species the structure is stiffened by regular, mineralized bodies called sclerites, spicules or ossicles. In many cases, the shape, size and ornamentation of the sclerites are essential in identifying the species.

Feeders and pumpers

The more obvious polyps in a colony are feeding polyps called autozooids. They have eight featherlike tentacles, instead of the smooth tentacles of hard corals. The tentacles bear stinging cells with which they capture their prey, small planktonic organisms. These are passed to the mouth then to the stomach, which is divided into eight chambers. There are also polyps with very poorly developed tentacles and whose stomachs are not divided into chambers. These are called siphonozooids. They pump water into the colony, primarily for oxygenation, but in many species they also play an important role in maintaining internal pressure, to keep the colony relatively stiff and erect.

Soft corals are generally found in shallow waters between low tide and about 160 feet (50 m), very few having successfully colonized deep waters. There are numerous species that favor dimly lit caves and overhangs, but others that need well-lit areas. They are particularly abundant at 33–100 feet (10–30 m) on coral reefs. Possibly due to the lack of need to deposit the mineralized skeleton, soft corals are often much faster growing than hard or stony corals that form reefs. In many cases, when reefs have been destroyed, soft corals are among the first large organisms to reappear. On occasion they appear in such densities that they greatly hinder the establishment of hard corals.

Soft corals rarely have encrusting growths of plants or animals, unlike many other reef organisms, in part because they secrete potent chemicals that inhibit the growth of marine organisms. For this reason, medical researchers eager to find possible tumor inhibitors have intensively studied soft corals.

Microalgal partners

Many species of soft corals have embedded within them extremely numerous tiny micro-algae, or *zoothanthellae*. These photosynthesize like all plants, and can release significant quantities of food into the host coral. In some species of *Xenia*, the zoothanthellae are the only source of food, the action of the polyps being solely aimed at pumping water. This is more common in the hard corals. Hard and soft corals also share the phenomenon of coral bleaching, where zoothanthellae are expelled under conditions of stress, particularly high temperatures. However, this phenomenon is less widespread than in hard corals, and perhaps less likely to lead to the ultimate death of the colony.

Some soft coral species are hermaphroditic (male and female organs in the same colony). Sexual reproduction is common, with the eggs and sperm being shed into the water, where fertilization takes place. The *planula* larvae that hatch from the eggs are radially symmetrical, ciliated and free swimming. After some days or weeks in the zooplankton the planulae settle and develop into polyps. Each polyp soon begins to bud new polyps, and a new colony is begun. Unlike hard corals, some species of which can remain as solitary polyps, soft corals are always group-living. Some species can also reproduce by fragmentation: drifted pieces of colonies, perhaps broken off during storms, can reattach elsewhere and continue to grow, so forming new colonies.

Varied and beautiful

Soft corals are often vividly colored. Shades of red, pink, purple, yellow, green and brown are common, often reflecting the colors of the zoothanthellae. They also come in a dazzling variety of shapes: encrusting forms, mushroom shaped, lobed, or branched to varying degrees of complexity. *Sarcophyton* species, such as the massive, gently lobed elephant's ear coral, *S. trocheliophorum*, are probably the largest, regularly reaching over 40 inches (1 m) across. Many soft corals can be mistaken for stony corals until a gentle waft of current reveals how soft and flexible they really are.

Most soft corals feed mainly at night, perhaps to avoid the action of feeding fish, which like to bite off the polyps. A number are conspicuously active during the day, however, including the genera *Xenia* and *Heteroxenia*. Many of these have polyps extended for 24 hours per day, except when disturbed. These open and close for feeding at typically 40 times per minute in rhythmic pulses. This creates regular waves of contractions that spread over the colony surface.

ELEPHANT'S EAR CORAL

PHYLUM	**Cnidaria**
CLASS	**Anthozoa**
ORDER	**Alcyonacea**
FAMILY	**Alcyoniidae**
GENUS AND SPECIES	***Sarcophyton trocheliophorum***

LENGTH
Polyps: ½ in. (1.3 cm); colonies: 40 in. (1 m)

DISTINCTIVE FEATURES:
Colony of polyps supported on short, massive stalklike structure, 5 to 10 large, waved lobes spread from stalk. Polyps long and tube-shaped.

DIET
Plankton; nutrient uptake from symbiotic zoothanthellae; direct nutrient uptake from sea water

BREEDING
External fertilization; breeding season varies geographically, often year-round; ciliated planula larvae spend days or weeks in plankton before settling

LIFE SPAN
Probably more than 10 years in wild

HABITAT
On rocks and other corals in open, sunny positions, from low water to 100 ft. (30 m)

DISTRIBUTION
Indian and Pacific Oceans; Red Sea

STATUS
Locally common

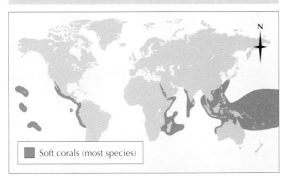

Soft corals (most species)

The polyps of these soft corals are visible as red or pink objects on the white structure, called the coenenchyme. Swimming among the corals are black-margin sweepers, Pempheris mangula, and copper sweepers, P. oualensis.

SOFT-SHELLED TURTLE

The spiny soft-shelled turtle is widespread across North America, from the St. Lawrence River to the Rocky Mountains. This is the eastern subspecies, A. s. spiniferus.

THE SOFT-SHELLED TURTLES are unlike other turtles because their shells are covered by a layer of leathery skin instead of horny plates. The shell is flat and almost circular, which gives some of them the alternative name of flapjack turtle. The jaws are hidden under fleshy lips and the snout is drawn out into a tubelike proboscis. The fourth and fifth toes are elongated, clawless and webbed to form paddles.

Soft-shelled turtles are found in many parts of Africa, Asia and America, and fossils of one genus, *Trionyx*, have been found in many parts of Europe. There are three similar species of soft-shelled turtle in North America: the Florida softshell, the spiny softshell and the smooth softshell. It is often difficult to tell these apart in places where their distributions overlap.

Some of the Asian soft-shelled turtles, such as *Lissemys punctata* of India, are called flap-shelled turtles or flapshells. They have hinged

flaps of skin at each end of the plastron, the underside of the shell, which completely close up the shell when the head and limbs are withdrawn. The Indian flapshell is the smallest soft-shelled turtle, along with the Chinese soft-shelled turtle. Neither of these species ever have a shell that is more than 11 inches (28 cm) long. The largest species is the Asian giant soft-shelled turtle, in which the shell may grow to a length of more than 4 feet (1.2 m).

Although some species are relatively abundant, many more species are undergoing dramatic declines in numbers, especially in Asia, where markets in turtles have opened up for food and traditional medicine, as well as the exotic pet trade. Many species are heading towards extinction.

Supplementary breathing

Soft-shelled turtles usually live in fresh water, but the rare and endangered long-headed turtle, *Chitra indica*, can live on the sea shore. They are extremely active, perhaps as a result of their lightweight shells. They have been reported to swim at 10 miles per hour (16 km/h) and they can also run fast on land. These turtles must be handled with care as they can strike rapidly, sometimes causing nasty wounds.

Although capable of bursts of great activity, soft-shelled turtles spend much of their time lying almost submerged in the muddy bottoms of lakes and rivers. They can stay underwater for long periods by stretching their long necks until the tubelike nostrils break the surface of the water, rather like a submarine's snorkel. They can also stay under by using a supplementary method of breathing: using both mouth and rectum as gills, they extract oxygen from the water that is circulated through them.

Powerful jaws

Like other freshwater turtles, such as the snapping turtles, the soft-shelled turtles live on a variety of food, including plants, carrion and live animals. The main food of the spiny soft-shelled turtle is crayfish and aquatic insects, so it is doubtful whether it ever competes with anglers, as is sometimes asserted. The jaws of these turtles appear deceptively soft because they are hidden by fleshy lips. They are, in fact, very

SOFT-SHELLED TURTLES

CLASS	**Reptilia**
ORDER	**Testudines**
FAMILY	**Trionychidae**

GENUS AND SPECIES **Spiny soft-shelled turtle,** *Apalone spinifera;* **smooth soft-shelled turtle,** *A. mutica;* **Florida soft-shelled turtle,** *A. ferox;* **Asian giant soft-shelled turtle,** *Pelochelys bibroni;* **others**

ALTERNATIVE NAMES
Pancake turtle; softshell

LENGTH
Most species: shell length 20–30 in. (50–80 cm). *P. bibroni*: **shell length 48 in. (1.2 m)**

DISTINCTIVE FEATURES
Shell covered with leathery skin; neck relatively long; snout elongated into probiscis, with nostrils at tip

DIET
Invertebrates and small fish mostly; various dead or living animal and plant material

BREEDING
A. spinifera: **age at first breeding: 8–10 years; breeding season: April–October; number of eggs: 5 to 38; breeding interval: 6 months**

LIFE SPAN
Up to 50 years

HABITAT
Fresh water, occasionally brackish. Large, slow-moving rivers often preferred.

DISTRIBUTION
North America; most of Africa except deserts; Asia from Turkey to Japan and New Guinea. Introduced to Hawaii.

STATUS
Vulnerable (many species). Endangered (Burmese peacock softshell, *Nilssonia formosa;* **4 others). Critically endangered (***Aspideretes nigricans;* **4 others).**

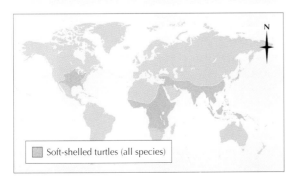

Soft-shelled turtles (all species)

strong. In some species the jaws have a sharp cutting edge suitable for dealing with fish, whereas others have jaws with surfaces better suited for crushing the shells of mollusks.

Nesting takes place in spring or summer in North America. The females leave the water to dig nests on land. These nests are similar to those of many other turtles: they are flask-shaped holes dug as far as the hind legs reach, which is up to 1 foot (30 cm) deep. From 10 to 25 round white eggs are deposited in the nest and covered over with sand or soil. A female may sometimes lay more than one clutch a year, and occasionally eggs laid in the fall may not hatch until the following spring.

Sacred turtles

Some species of soft-shelled turtles are captured in large numbers for food, but others are protected for religious reasons. There are several ponds containing sacred turtles in parts of Asia. These animals are often extremely tame, and will come to be fed by hand when called. The most unusual sacred turtles are at Chittagong in Bangladesh. There are between 150 and 200 of them in tanks of fresh water. They were first scientifically described in 1875, and are the only surviving individuals of the black soft-shelled turtle, *Aspideretes nigricans*, in the world. The original distribution of the species is not known, and it is critically endangered. In parts of West Africa, the Senegal flapshell turtle, *Cyclanorbis senegalensis*, is kept for a more mundane purpose. It is placed in wells to eat any refuse that falls in.

The long neck and proboscis of the Florida soft-shelled turtle allow it to sit on the bottom of shallow ponds and swamps while reaching the surface with its nostrils. From there it can ambush prey.

SOLE

Although drab in color when seen for sale in the market, in life the sole can adopt the color and patterning of its surroundings. The thin covering of sand on this sole enhances its camouflage.

THE COMMON SOLE IS considered by some connoisseurs to have the best flavor of all fish, but only after it has been dead for two to three days. Soles are tongue-shaped flatfish in which the mouth is on the lower side of the head. The head projects forward of the mouth in a smoothly rounded curve. Sole orient their flattened bodies to lie on their left sides. The dorsal fin starts on the front of the head and continues around the margin of the body to join the tail fin, the anal fin being nearly as long but starting behind the gill cover. Dorsal, tail and anal fins form a complete fringe. The underside of the head is covered with little white tendrils, crowded together.

Most soles are up to 1 foot (30 cm) long but some grow to be double this length. The usual color of the upper surface is yellow, grayish brown or dark brown, with well-spaced darker spots or blotches, or with dark bands. One of the more striking species is the naked sole, *Gymnachirus nudus*, of the American Atlantic coasts, which has a zebra pattern, with close-set, reddish brown crossbands.

Soles live mainly in shallow seas, on muddy or sandy bottoms, although a few species live in deep water, such as the deepwater sole, *Bathysolea profundicola*, which ranges between depths of 900 and 3,800 feet (270–1,160 m). Most live in warm seas but some species, including the common sole of Europe, range into temperate seas as far north as the Faeroe Islands. The hog-

choker, *Trinectes maculatus*, of America, which is 6 inches (15 cm) long with crossbands on the upper surface and spots below, lives in the sea from Carolina to Panama, and sometimes enters fresh water.

Timid behavior

Soles lie on the sea bed, more or less buried in the sand during the day. They bury themselves with a strong undulatory movement of the body, which digs a shallow trough. The sole's movement also throws up sand that settles on its body, partially hiding it.

Soles normally become active at night, which is their main feeding period. They are also active by day when the skies are overcast or when the water becomes murky. They then creep slowly along the bottom using the ends of the fin-rays with which their bodies are fringed.

Arrayed senses

A sole searching for food raises its head slightly upward and sideways, patting the sand from time to time with the underside of its head. As both of the eyes are on the upper (right) side of the body, they seem well adapted to detecting predators descending from above, but poorly adapted to finding the bottom-dwelling animals that the sole itself preys upon. To hunt, the fish seems to rely instead on the senses of touch and smell. A sole will stop and go back to examine any small object its undersurface has touched, feeling it with the lower surface of its head. The little white tendrils are sensitive and are probably organs of touch, possibly also of taste. The tubular nostril on the underside suggests that it may also use smell in searching for food. One nostril of the sand sole, *Pegusa lascaris*, is surrounded by a rosette of swollen skin, which may be associated with its dual touch and smell method of sensing prey.

The sole feeds entirely on bottom-living animals and seems incapable of catching, or even noticing, anything swimming just off the bottom. Its upper jaw and teeth are feeble and most of its mouth is on the undersurface of the head. Small crustaceans, worms and mollusks make up the main part of its diet, and to these are added brittlestars and small bottom-living, fish such as gobies. Prey is not seized in the mouth until the sole has been able to cover all or part

COMMON SOLE

CLASS	**Osteichthyes**
ORDER	**Pleuronectiformes**
FAMILY	**Soleidae**
GENUS AND SPECIES	***Solea solea***

ALTERNATIVE NAME
Dover sole

WEIGHT
Up to 6½ lb. (3 kg)

LENGTH
Up to 28 in. (70 cm)

DISTINCTIVE FEATURES
Flatfish: body flattened laterally, left side faces downwards; both eyes on right (upper side), mouth on left; small head; dorsal fin starts in front of upper eye; upper pectoral fin has black spot and is slightly larger than that on lower (blind) side; dorsal, anal and caudal fins form continuous fringe around body; white tendrils in fringe around head

DIET
Worms, mollusks and small crustaceans

BREEDING
Age at first breeding: 4 years; breeding season: January–April (south), April–June (north); number of eggs: 500,000

LIFE SPAN
Up to 20 years

HABITAT
Sandy and muddy seabeds in shallow waters

DISTRIBUTION
Eastern Atlantic from southern Norway to Cape Verde Islands and Senegal; North Sea, Mediterranean, Sea of Marmara, the Bosphorous and Black Sea

STATUS
Common

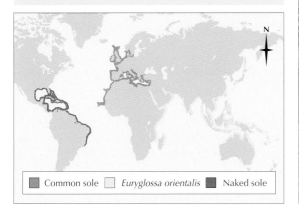

Common sole   *Euryglossa orientalis*   Naked sole

of it with the front part of its head. Young soles feed on fish larvae and smaller crustaceans such as copepods.

Flattened during development

The common sole and some other species move into deeper water for the winter but migrate back to shallow inshore waters in spring and early summer to spawn. The eggs, ⅟₂₀ inch (0.2 mm) in diameter, float well off the bottom. Each mature female lays about half a million eggs. The larvae are ⅛ inch (0.5 mm) long when hatched and, as in other flatfish, have the symmetrical bodies of unspecialized fish. As they grow they come to lie on the left side, and the left eye migrates to the right side, as in plaice. By the time they are ¾ inch (3 mm) long their bodies have become flattened and they have settled on the bottom.

Subterfuge and deceit

When the sole is dead, its upper surface is a uniform sepia brown. In life most soles harmonize with the seabed on which they happen to be lying, their camouflage being helped by blotches, spots and bars, which further break up the outline of the body. The fact that a mature female lays over 500,000 eggs as compared with millions recorded for other fish is a good indication that soles enjoy a low rate of attack. If their camouflage does not work, the sole can fall back on another defence tactic, using the black patch on its pectoral fin. The poisonous weever fish, which also buries itself in the sand, has a black patch on its dorsal fin, which it raises when a predator approaches. This is probably a warning signal to attackers that the fish is poisonous. The sole also raises its black-spotted fin, so perhaps this is a case of mimicry, the sole benefiting from being mistaken for a weever fish.

In Europe the common sole is still marketed by the fishing industry as the Dover sole.

Index

Page numbers in *italics* refer to picture captions.
Index entries in **bold** refer to guidepost or biome and habitat articles.

Page numbers in *italics* refer to picture captions. Index entries in **bold** refer to guidepost or biome and habitat articles.